T0258891

Reliability Centered Maintenance-Reengineered

Practical Optimization of the RCM Process with RCM-R®

Reliability Centered Maintenance-Reengineered

Practical Optimization of the RCM Process with RCM-R®

By

Jesús R. Sifonte, James V. Reyes-Picknell

CRC Press
Taylor & Francis Group
Boca Raton London New York

CRC Press is an imprint of the
Taylor & Francis Group, an **informa** business

CRC Press
Taylor & Francis Group
6000 Broken Sound Parkway NW, Suite 300
Boca Raton, FL 33487-2742

© 2017 by Taylor & Francis Group, LLC
CRC Press is an imprint of Taylor & Francis Group, an Informa business

No claim to original U.S. Government works

International Standard Book Number-13: 978-1-4987-8517-4 (Hardback)

Library of Congress Cataloging-in-Publication Data

Names: Sifonte, Jesus, author. | Reyes-Picknell, James V., author.
Title: Reliability centered maintenance-reengineered (RCM-R) : practical
optimization of the RCM process / Jesus Sifonte and James V.
Reyes-Picknell.
Description: Boca Raton, FL : CRC Press, 2017.
Identifiers: LCCN 2016046134 | ISBN 9781498785174 (hardback : alk. paper)
Subjects: LCSH: Plant maintenance--Management. | Maintenance--Management. |
Reliability (Engineering) | Maintainability (Engineering)
Classification: LCC TS192 .S59 2017 | DDC 620/.00452--dc23
LC record available at https://lccn.loc.gov/2016046134

Visit the Taylor & Francis Web site at
http://www.taylorandfrancis.com

and the CRC Press Web site at
http://www.crcpress.com

Contents

Foreword

This book by James Reyes-Picknell and Jesús R. Sifonte is a welcome addition to the literature on reliability centered maintenance (RCM). Over the last decade, the workplace has changed, and RCM has evolved to meet the needs of today's companies. Although papers have appeared charting this evolution, the book represents an opportunity to reflect on and consolidate the findings. This is not a backward-looking volume, however. Far from it. Rather, its cutting-edge analysis points to the continued relevance of RCM well into the future.

Reyes-Picknell and Sifonte are extremely well positioned to tackle the project—with strong backgrounds in both theory and practice. They begin the book with an explanation of the value of RCM in the current context. Then, in Chapter 3, they hint at the future with an introduction of the changes involved in their formulation of RCM-R® (reliability centered maintenance-reengineered). Of course, RCM-R® is not simply a theory, and the following chapters go on to explain its practical application—supplemented by numerous very helpful examples, along with figures highlighting the main points.

To put it simply, RCM-R® takes RCM a step further by making it more evidence based where data are available. In general, RCM-R® requires operational, technical, reliability, maintenance-related, failure, material, financial, safety, and environmental data to be analyzed for decision-making purposes. The effort to take RCM a step further is, in fact, an overriding theme of the book. A particularly valuable example is Chapter 9 on condition-based maintenance techniques, where the authors, along with several colleagues, contribute detailed insights into the condition monitoring technologies of vibration analysis (Jesús Sifonte), infrared thermography (Wayne Ruddock), lubrication and oil analysis (Mark Barnes), ultrasound (Allan Rienstra), and nondestructive testing (Jeff Smith). In Chapter 12, they extend RCM into the provisioning of spare parts, including a section on slow-moving capital spares written by Neil Montgomery.

Andrew K. S. Jardine
University of Toronto

Introduction

WHY RCM-R®?

RCM was successfully developed during the 1970s while the aviation industry was facing important challenges. High maintenance costs, frequent undesirable component failures, and the safe increase in passenger capacity of aircraft designed to meet Federal Aviation Agency requirements were some of the struggles the aviation industry overcame with the development of its novel process to determine failure consequence management policies for aircraft nonstructural components. Later on, the so-called reliability centered maintenance process was introduced with success in the mining and nuclear industries. The development of an international asset management standard (ISO 55000) in 2014 highlighting the need for realizing optimum value from physical assets has evidenced the necessity of a qualified process for asset and process risk assessment and management. RCM, as conceived in the Society of Automotive Engineers JA1011 and 1012 standards, has again emerged as a four-decades-old proven process to face today's challenges. A variety of RCM processes are used in almost every industrial sector today. There are some impressing RCM success stories, and there is also a lot of frustration when the expected results are not achieved.

A Fortune 100 manufacturing company with high quality standards had already implemented some predictive and precision maintenance technique efforts at a local facility. Global management decided that the local facility needed to implement RCM for its critical assets to improve the plant's throughput. The newly hired reliability engineer was required to lead failure modes and effects analysis for the facility's critical assets with a group of maintenance technicians. Their analysis yielded new maintenance procedures, which were implemented right after management approval. Unfortunately, after 3 years, the company had not obtained the expected results from their RCM effort, and they looked for external help to understand what was being done wrongly. A qualified RCM-R®

facilitator noted the following fundamental pitfalls when auditing the project:

- No use of experienced RCM facilitator
- No multidisciplinary team approach used
- Poor failure mode causation
- Only maintenance tasks recommended (no one-time changes)
- Too few C over T type tasks recommended
- Only primary functions evaluated

Regrettably, the success of the effort depended on only one person, who lacked both condition monitoring and reliability engineering experience, resulting in less than optimum results. The use of abbreviated computerized methods relying only on the input of maintenance and the lack of knowledge of predictive and precision maintenance hindered optimum results from happening.

Another facility of the same organization learned the lesson from their peers, and they were able to run a very fine project with astonishing results. Their work included consequence management policies resulting in maintenance tasks performed by technicians and operators, changes to operating procedures, some minor machine redesign, and spare parts inventory level modifications. They used a trained multidisciplinary team together with an experienced RCM-R® facilitator. The team also incorporated the use of statistical failure analysis to fine-tune their consequence management policy selection. They found that the statistical analysis of the critical failure events documented in their corrective maintenance work orders often revealed something different from the assumed physics of failure. Reliability, availability, and maintainability analyses helped the team to quantitatively establish the system's current state as the base for improvement. The project results reported after a year of implementation showed a 63% reduction in preventive maintenance man-hours and over 50% reduction in overall corrective maintenance costs. After some years of their first project implementation, the company continues to monitor the asset's performance with reliability, availability, and maintainability analysis. They have analyzed some new failure events not considered during the RCM analysis. The asset has also undergone some new component modifications, changing its inherent reliability, while some new predictive maintenance instrumentation has been acquired. Continuous improvement is a cornerstone for a sustainable asset management effort. Assets undergo operating context variation, physical modifications, or even changes in their criticality over their operating life.

RCM-R® is an updated failure consequence management policy development process incorporating several techniques into the well-proven SAE JA1011 RCM method to ensure that knowledge from asset maintenance and operation experts is efficiently used to attain the organization's goals. RCM-R® is a tool to facilitate corporate learning and knowledge transfer from those who are most experienced to those who are not. Ensuring asset data integrity is an essential aspect of RCM and of the whole asset management decision-making process along all the life cycle stages of an asset. The process develops assets and failure events criticality ranking matrixes per International Standards Organization standards, enabling organizations to concentrate their scarce resources on vital aspects of the business that impact their goals while filtering out nonessential matters. RCM-R® feeds information on what is actually critical to digitize and specifically, what data needs to be collected and monitored. This is very important as we move toward the "internet of things" whereby virtually everything can talk to everything else. RCM-R® helps us become better digitally informed, rather than digitally distracted! RCM-R® enhances the capability of the traditional analysis with the introduction of prework stages required for getting the right facts and data, aiding the team to prioritize analysis efforts and eventual actions. It also improves the effectiveness of the selected tasks by the use of evidence-based maintenance founded on statistical analysis of actual asset failure events. The incorporation of the quantitative analysis sets the grounds for calculating the asset's inherent reliability and availability. Finally, the living RCM-R® process becomes part of a day-to-day culture in which practitioners strive to continually improve the plant's performance for attaining financial, safety, and environment-related company goals.

1

Asset Management

Reliability centered maintenance (RCM) is the most successful method we have for developing failure management policies with the aim of sustaining the functional performance of our physical assets. Despite its name, it goes well beyond maintenance to include operational, engineering, procedural, process, and training outcomes. It can touch on many aspects of our businesses if we use physical assets—and these days, just about everyone does. Since RCM has such broad implications, where does it fit into modern business management systems?

RCM is a key element of good asset management as we know it today, because both have similar desired outcomes. It belongs in any good systemic asset management system in modern businesses. Let's look at why and how.

Asset management has been around from prehistoric days, when people first thought of using tools and weapons. Since then, we have been making decisions about what we need tools to help us with, what tools can be used to do it, their form and functionality, how they should be maintained (repaired or replaced), how to improve on them, and how to dispose of them. As our physical assets became more complex, the decisions and follow-on actions got more involved. For example, railroad systems are more complex to manage than a fleet of stagecoaches on open prairie lands. People began to specialize in aspects of asset management: for example, making crockery, bricks, and so on, design and construction (building the pyramids), installation (putting horseshoes on horses), repair of metal tools and pots (blacksmiths), logistics (transport of building materials and other goods), manufacturing in larger quantities (the Industrial Revolution), research and development of new methods (alchemy, chemistry, nuclear physics), controls, information management, and so on. Specialization has enabled humans to do more and do it better. Few today are "jacks of all trades"—we tend to be masters of one. Working together, all these masters can produce amazing results. This has resulted in a world full of physical assets that we depend on all the time, and sometimes, due to our specialization and divisions among who does what, we don't even know where those dependencies lie—till something breaks. When something fails, we feel the sting of our reliance on technology. The next step in our evolutionary path may well be that we need to get better at managing complexity and interconnectedness. Asset management is just such a discipline.

Getting value from physical assets involves a lot of separate but related activities. Marketing forecasts future demand, designers determine how best to provide a supply to meet the demand, engineers design and build it, finance pays for it, operators get it going and doing what it is needed to do, and then maintainers keep it going. Each of these disciplines is quite complex, and whole careers can be spent in just one area. Getting the whole thing to work to our greatest advantage takes an integrated and holistic effort. The asset life cycle is complex, yet we need to manage that life cycle so that we get the performance we want, at a price we can afford, with tolerable risks to safety, the environment, and our business. Without consideration of the whole cycle, we will get poor performance or capabilities we can't use. To a degree, supply and demand for capabilities takes care of this, but each element is optimized independently of the others. The whole isn't optimized as one system. Asset management addresses that shortfall. It is the discipline of managing these physical assets—more importantly, what they do for us throughout their life cycle. RCM is a tool that good asset management can use to help achieve that, but today, only a few industries have leveraged it fully.

Some industries have a natural need to be better than others at managing assets. Utilities, transportation networks, airlines, and the military all tend to have a lot of pressure to manage assets well because of safety, environmental factors, high cost, or other reasons. Interestingly, those industries were the earliest adopters of RCM. To keep costs down for rate-payers, municipalities have gradually become better at planning and forecasting future needs, operating and managing maintenance for their assets (streets, traffic controls, office buildings, schools, hospitals, etc.), and many still have a long way to go. In the public interest, many municipal jurisdictions are required to have asset management plans of one sort or another. The earliest to do this didn't always refer to any particular specification or standard, so results have been mixed.

In some places, the cost of making mistakes has historically been quite high. New Zealand, Australia, and South Africa are somewhat remote and sparsely populated. They lack the scale of industrial base that Europe and North America have. Consequently, they've had to manage their assets well or wait a long time when something goes wrong for parts and materials to restore whatever has failed. They have also become very creative and have led the world with some innovative approaches to management of physical assets. The first formalization of asset management as a discipline arose in New Zealand and Australia. Today, we see a lot of interest in the field of asset management in developing countries. They have similar challenges to those faced in Australia and New Zealand, and they have the benefit of asset management thinking, which evolved to face those challenges. As they adopt these practices, they accelerate their development and position themselves well to compete globally, and not only on the basis of lower labor costs. We in the more developed world would be wise to take heed.

The United Kingdom has a large population in a small landmass. It has an infrastructure that blends very old historical structures with much newer ones. It has a complex road and rail system and dense distribution systems for electric power, water, gas, and telecommunications. As aging systems failed, they created a great deal of disruption, inconvenience, and even injuries and fatalities in some cases. Simply replacing aging assets is a costly option, and they knew they could manage them better. The need to manage this led to the early adoption of what was being pioneered in Australia and New Zealand regarding managing assets. In 2004, the United Kingdom produced the world's first "specification" for asset management—PAS 55-1 and 55-2. Those specifications were first put into use in gas, electric, and water utilities in the United Kingdom. They instilled a discipline around managing assets that began to show benefit and promise. Those specifications and their quick adoption by the UK network utilities got a lot of attention worldwide. By 2014, the International Standards Organization (ISO) had used UK specifications as a model to produce a series of three standards—ISO 55000, 55001, and 55002—outlining requirements for good asset management pertaining to our ubiquitous and increasingly complex physical assets. The standards are general in nature and apply to any sort of physical assets and the non-tangible assets associated with them, such as documentation, training programs, and so on. They are voluntary standards, so no one is forced to use them, but we can already see a trend toward their becoming mandatory.

ISO 55000 describes why we should be doing good asset management, 55001 specifies what must be done to achieve it, and 55002 describes a bit of "how to" along with helpful suggestions.

In the United Kingdom, and increasingly elsewhere, compliance with formal asset management standards is required. Municipalities that are required to have asset management plans are now turning to the available standards, because they provide a logical structure for those plans. Insurers are beginning to look for evidence of good risk management, and that comes with good asset management. If risks are poorly managed, premiums are higher. Given the high levels of insurance payouts due to disasters (many of which are arguably the result of poor asset management), the insurance industry is looking for ways to minimize its exposure. It does this by insisting on good practices that reduce its risks. Some have gone so far as to offer premium reductions to companies that have become certified against the formal standards and those that use methods such as RCM to help in managing risks.

The authors see the beginning of a trend. As societies become more litigious and more heavily regulated to drive safety and environmental compliance, they are increasingly looking for improved ways to do things. Whenever something goes wrong, we examine what went wrong—some of us do it with an eye to identifying improvements that need to be made; others do it to affix blame and begin legal proceedings. As more insurers and lawyers learn of these new standards, they will increasingly expect companies to follow them in order to demonstrate that they are doing their best as a form of due diligence. Before long, compliance will be expected. No doubt some jurisdictions or regulatory authorities will require compliance and possibly third party certification of compliance. Eventually, we will see negligent asset managers, their companies, and their executives being held to account by the courts for failure to follow good asset management practices. There are already cases of companies and their executives being jailed for negligence. What better way to defend yourself than to actually follow good practices and be able to document it?

We manage our businesses to deliver goods and services and to make a reasonable profit in doing so. A key outcome of good asset management is the management of risks to the business. There is a need to identify risks, decide what to do about them, do it, and document it so you can prove you've done it. Those risks can arise in a variety of ways throughout the life cycle of the asset, from concept through to disposal. A large part of the asset's life cycle is the operational phase, when the asset is being both used and maintained. The better the design of the asset, the more successful it will be in doing what its user wants of it. Early in the life cycle, anything that can improve the design without adding risks and costs is beneficial. Likewise, later in the life cycle, anything that can keep the asset operating reliably at optimum maintenance costs is also beneficial. It is in these two phases of the life cycle, design and operations, that RCM has the greatest beneficial impact.

RCM was first developed for application in the design phase for new aircraft. Its purpose was to improve flight safety (performance) at lower costs than if the industry had continued to use its historical practices. Chapter 2 discusses the history of RCM in more detail. RCM was really a very smart way to make the airline business more profitable while also making it safer.

The reliability of any physical asset is determined by the asset's design. Without changing aspects of the design, you cannot make something more reliable. However, you can make it worse quite easily. It is affected by how it is operated, its operating environment, and its maintenance or lack of it. In the aircraft industry, design, operations, and maintenance are tightly controlled, and aircraft generally do achieve high levels of reliability. Pilots know that aircraft cannot be flown safely outside of their specified operating envelope (altitude, angle of attack, speed, etc.). Doing so may result in failure of aircraft components or structure, a crash, and likely the loss of life. Consequently, they rarely exceed those operational limits. In aircraft fleets, therefore, with the exception of design changes and upgrades to equipment, reliability is largely in the hands of aircraft maintainers. RCM was therefore intended to address maintenance programs as a way to achieve reliability. The central focus for sustaining reliability was maintenance—hence the name *reliability centered maintenance*.

RCM embeds a method, failure modes and effects analysis (FMEA), to identify design flaws, weaknesses in training or skills, and operating practices as well as normal failures that should be expected in the course of operations. When RCM is applied at the design stage of any system's life cycle, it is used to help identify design enhancements.

When it is applied to systems already in operation, the opportunity to influence design is past. Nevertheless, even in operational systems, RCM may identify design flaws to be managed with maintenance. However, if this is not possible, users may find themselves in costly redesign scenarios. To avoid those situations, RCM is best applied as part of design efforts.

RCM has also found application outside of the aircraft industry. In nuclear power and military systems, it is used much as it is in aircraft—in the design phase. In other applications, the consequences of failures are not usually so catastrophic, yet RCM still has a role to play, and it has been applied in a wide variety of industries. What is different, however, is that most other industries do not apply it in the design phase. Let's look at why that happens and what occurs as a result.

Applying RCM in the design phase entails an additional investment of capital funding. Most organizations do their best to minimize upfront capital investment as a way of minimizing financial risk in new projects. Unfortunately, that approach is somewhat misguided. Low cost often results in poor quality and disappointing results. Similarly, when it comes to RCM, upfront savings often produce sustained poor performance in the long term—exactly what you don't want! Indeed, money is saved in the short term, but there is a substantial risk that more could have been earned and saved on an ongoing basis if RCM had been applied early.

Without RCM at the design stage, the design may not be as able to meet its functional requirements as well as expected. Also, the maintenance program is unlikely to be well suited to the asset in its operating context—something that is often unique to each business. The result of this is that new projects experience problems at start-up (teething problems) and can take a long time to ramp up to full production levels. That long period of low output is costly in terms of revenue lost. The teething problems often show up in the form of unexpected equipment failures. When that happens, the needed parts, tools, maintenance planning, and even workforce skills are often found lacking. This happens because the need wasn't forecast early enough, and so the necessities were not provided for in a timely manner. Yet, the failure is highly likely something that could and would have been forecast had RCM been applied. In terms of asset management, failures are the result of risks that were poorly managed.

In many industrial environments, ramping a new operation up to full capacity can take months, in some cases a year or more. If new cars couldn't be used to full capacity for months or years, we wouldn't buy them. Imagine a new airplane that must fly half full for its first few months of operations or a new military weapon system that won't hit its targets till many trials have been run. Such low levels of performance in a new system are not acceptable to most of us, and we certainly wouldn't want to pay for them, yet we do seem to tolerate them in industrial systems. Cars are built in the millions; there is a lot of operational history and experience guiding design, run-in practices and periods, and maintenance. Aircraft and the military don't have such large numbers of identical assets in use. Their level of operational experience with systems is lower, so they rely on RCM. Elsewhere in industry, we often have neither the extensive field experience nor the RCM analysis. Large-scale industrial facilities are not replicated in large numbers. There simply is no accumulated body of experience and knowledge such as we have in the automotive industry. Since we don't have that, we need something else. RCM provides that opportunity to do the right things to manage those assets as they go into and continue in service.

There are those who argue that RCM is labor intensive and expensive and therefore not worth doing. They have probably never experienced the benefits, or they wouldn't say that. Let's look at costs first.

The cost of an RCM program for a new design should be on the order of 2%–3% of the capital cost of the new asset. Extended warranty costs on the various components, equipment, and subsystems in a new plant can easily exceed that amount. Arguably, extending warranty coverage doesn't really buy you much, if anything. It is essentially a form of gambling, and you are betting against yourself. You only get a payback if things fail for you! Why not take every step you can toward avoiding those failures and being ready for those that cannot be avoided?

Maintaining validity of warranties can also be a costly administrative and operational headache. Warranties come with conditions. Usually, you must follow the manufacturer's recommended maintenance program and stick strictly within their operating parameters, which may even include a lengthy run-in period. Considering that most manufacturers of industrial systems actually produce a fairly limited finite number of those systems, that they are seldom identical, and that the manufacturers usually have little, if any, operating experience with their own systems, are their recommended practices really to be trusted? The authors' experience with RCM reveals that manufacturers' recommended practices are frequently very poorly suited and often downright harmful to the assets. We don't believe that there is malicious intent, but it is those same manufacturers who benefit from sales of parts and aftermarket services if you get into trouble with their products. They have little incentive to apply RCM to their own products and then to tailor their analyses to their customers' operating contexts.

Now, let's look at RCM's benefits. In the aircraft industry, as discussed in Chapter 2, there have been huge improvements in operational safety and cost performance. Arguably, the industry would be much smaller than it is today without RCM. The authors have applied RCM in a variety of diverse applications with great success. In most cases, the effort paid for itself long before the entire program of analysis was even completed. In a few cases, it was paid for many times over in the very first pilot project! Of course, those exact benefits are difficult to forecast before you actually do the analysis work. It's a bit of a chicken-and-egg situation—which comes first? Do we take it on faith that there will be a payback worthy of the investment, or do we wait until later to see what trouble we might get into and then turn to RCM as a fix? Sadly, despite a huge body of evidence that RCM is worth doing, most companies opt for the latter. Short-term cost focused thinking prevails! So, what pain does it take to get companies to adopt RCM?

Consider the story of a frog in a pot of water over a fire. As the water warms up, the frog adapts to the warmer temperature, but it keeps getting hotter. Eventually, it gets too hot, but the frog is weakened by the heat and can no longer jump out. It ends up in a boil, all because it didn't leap out sooner while the discomfort was only minor. People are funny beings, sometimes a bit like the hapless frog. We are very capable of getting comfortable with discomfort. We tolerate situations that are uncomfortable to us for a long time till they get so bad, so uncomfortable, that we eventually decide to do something about them. We do this in all aspects of our lives, and at work in industry we are no different. Of course, in industry, that behavior and the pain it causes are on an industrial scale!

In the authors' experience, many industrial applications of RCM occur during the operational phase of the life cycle, and they are usually preceded by a long history of unacceptable performance due to unreliability of the physical assets. In those cases, RCM has been identified as a solution to what has often already become an intractable problem. That problem, or series of problems, has been bleeding cash from the operation. Revenues are falling short even in good markets, because production can't keep up. They can't keep the production lines up and running at capacity and reliably. Breakdowns are disruptive to production and impact on product quality, cost of raw materials in process, scrap, and so on. Frequent breakdowns (often coupled with poor planning and materials support) usually lead to expensive repairs, excessive consumption of parts, and overtime costs. In some cases, the equipment failures may have resulted in situations that gave rise to safety risks or accidents. Fines and lawsuits, even injuries and fatalities have resulted. In other cases, the failures may have resulted in emissions that exceeded allowable limits or leaks of hazardous substances. Again, fines, loss of license to operate, and negative public opinion have resulted. Poor operational reliability makes business forecasting a challenge, and conservative forecasts to compensate for it are generally not considered to be acceptable solutions. Poor forecasting and the inability to meet forecasts can result in devastating stock market performance. The high costs of keeping things running result in budget overruns, later followed by belt tightening. In maintenance and operations, that can translate into less training or less proactive maintenance. One keeps skills depressed, while the other almost guarantees that more failures will soon follow. It is a negative downward spiral that is challenging to arrest and reverse. If not corrected, this situation will probably lead to bankruptcy and closure of the business. In desperation, a fix is needed.

Believe it or not, these scenarios play out often around the world. Sometimes RCM can help, and sometimes the situation is too far gone, so we lose another frog. RCM, despite all of its benefits, is not a quick fix or a cure-all. It also doesn't work in isolation. You cannot just RCM your way out of a bad situation. It is a tool in a much broader asset management system—formal or informal. That system is the collection of activities your organization does to manage aspects of physical assets and their life cycle, whether they are managed as a holistic system or not. Of course, it's best to manage them together in a cohesive, strategic manner. That's what the asset management standards can help with. But if you don't have that in place, you will need to start putting aspects of it in place to get results and get your frog out of the hot water.

If an organization is in this situation of being in trouble and needing to do something differently, it probably has several things going wrong. For starters, its preventive maintenance (PM) program, assuming it has one, may not be followed very closely, or it may comprise PM tasks that are ineffective. RCM can deal with the latter but not the former. PM programs comprise predetermined work (tasks) done at specified intervals, either time or equipment usage based. For example, changing oil in a car every 10,000 km is a scheduled PM task. If you don't do it or fail to do it in a timely manner, then you put your car at risk. Getting the work done on schedule requires some discipline to follow the schedule. But doing so requires a few other "conditions" to exist.

In organizations where we find poor compliance with scheduled PMs, we often find a number of other problems:

- Too much breakdown work (a symptom): demonstrates that PMs don't work or are not being applied.

- Reactive culture: breakdowns get most of the attention. Equipment failures are driving the schedule, and the organization is failing to manage those failures. PMs (if used) get lower work priority and are often done very late (if at all).
- Poor planning: work, when done, is poorly planned or not planned at all. Lack of job estimates renders scheduling ineffective, and lack of preparation for the work (i.e., lack of parts, tools, etc.) ensures the job will take a long time.
- Poor communication among operations, maintenance planning, work crew supervision, and materials management.
- Maintenance likely has a poor reputation among operations personnel: they are failing to get work done when promised (if they promise at all) and often need to do work over.
- Maintenance costs are high.
- Operations are unstable: unreliable plant and equipment creates high variability in outputs from production lines, often accompanied by difficulties in maintaining quality levels. Depending on the nature of the operation, this can result in a lot of rework, off-spec product, or scrap.

RCM can do a great deal, but it is focused on "what" you do, not "how" you do it or "how" you get it done. RCM works best in an environment where precision maintenance is the norm and where management systems are able to sustain high levels of production with low levels of variability. If an organization cannot do this, it has much more work to do to benefit as much as possible from RCM and to get even close to good asset management practices.

Asset management is a broad and holistic discipline that includes the management of maintenance and uses RCM as a primary tool to drive maintenance and engineering decisions. It is difficult, if not impossible, to excel at maintenance management and asset management without RCM.

Here, we leave the field of maintenance management and asset management to other work such as "Uptime"[1] and the new international standards,[2] ISO 55000, 55001, and 55002. The remainder of this book will focus on RCM and specifically enhancements to the basic method that we call RCM-R® (RCM-Reengineered).

REFERENCES

1. John D Campbell and James V Reyes-Picknell. *Uptime: Strategies for Excellence in Maintenance Management*, 3rd edn, 2015, Productivity Press, New York, USA.
2. International Standards Organization, ISO 55000. *Asset management: Overview, principles and terminology*; ISO 55001. *Asset management: Management Systems—Requirements*; and ISO 55002. *Asset management: Management systems—Guidelines for the application of ISO 55001*, January 2014, International Standards Organization, Geneva, Switzerland.

2

The History of RCM and Its Relevance in Today's Industry

Is reliability centered maintenance (RCM) worthwhile? This is a tough question to answer if you do not know where RCM comes from. Today's most recommended methodology for choosing maintenance plans for critical assets didn't just appear. It has a complex background.

Today, industry faces a lot of challenges that were not even imagined some time ago. Could the employees of the remote small-town shoe factory imagine their product, one that they wore proudly from a company they all loved, being replaced by a foreign shoe, landing their company in bankruptcy? Their loyalty to their company and product wasn't enough. Those country boys now wear shoes made in China and bought in a Mega Store for half the price. For example, why do rum makers in Puerto Rico, the biggest rum producer in the world, need to buy sugar cane harvested abroad? Was Puerto Rico not the sixth largest sugar cane producer in the world with 51 sugar mills some five decades ago? Today, none of those remain. Industry is volatile; change is inevitable. No one can count on future success based only on brilliant past performance. Competitiveness is much more complicated today than it was 50 years ago.

In the past, competitors saw each other face to face often. They each knew what the others had to offer. But today, your most threatening competitor could be located in a distant continent and bringing goods to your customers, who find more value in getting them from him. Sadly, this is all too familiar. History repeats itself. We develop a product. It is produced, used, improved, and eventually mass produced. Eventually, our processes are rendered inefficient and need to be reengineered to survive the challenges posed by internal or external threats. Poor practices, competition, operational costs, tax policy changes, and so on all conspire against us. We respond or we fail. This happened in the aviation industry many years ago.

RCM DEVELOPMENT

The first commercial flight is credited to pilot Tony Jannus, who transported passenger Abraham Pheil from St Petersburg to Tampa at an altitude of 15 ft across open waters. The 21 mile flight lasted 23 min, and the passenger paid $400.00 for the ride. The era of civil aviation was about to begin. But, it was not till 1926 that the first true commercial flight took place. Postal mail was delivered by a United Airlines aircraft for the first time ever. In the meantime, Mr. Frederick Handley Page, a British aircraft manufacturer and designer, led a committee that evaluated the needs of civil aviation. They recognized that civil and military aviation requirements differed. The result of his investigation was the creation of the Air Registration Board (ARB) in 1937. ARB was responsible for issuance and renewal of certificates of airworthiness for commercial airlines in England. It also approved maintenance schedules based on manufacturer's recommendation. Aviation companies needed to comply with such maintenance plans to retain their airworthiness certifications. Almost all recommended maintenance tasks consisted of overhaul of parts before they reached their end of useful "life," expressed in operating hours. This protected the industry from negligent carriers who did not comply with requirements for certification. In those days, most engineers responsible for maintenance believed all parts followed the "bathtub curve" pattern of failure shown in Figure 2.1. They realized that care must be taken when doing maintenance, because premature failure can occur right after an overhaul. They believed

FIGURE 2.1
The bathtub curve, representing the old perception of a part's unique failure pattern.

that following the wear-in phase age, all parts would experience a reasonably long period of operation with a very low failure rate called *useful life*. This was then followed by a period of increasing failure rate in which wear- or burn-out failures would occur. As a maintenance strategy, care must be taken to ensure parts were replaced just before that wear-out period, as recommended by the approved maintenance schedules. There was a general belief that the amount (hours) of maintenance effort applied to an aircraft had a direct relation with its reliability and durability.

By the 1950s, commercial airplane size had increased extensively compared with the aircraft of the 1930s. Commercial airlines complained about the airworthiness process and expressed suspicion that many maintenance tasks were not necessary. They had observed that many overhauled parts were still working well when maintenance was carried out. Furthermore, they noticed unexpected failures after the overhauls, inferring that excessive repair work induced excessive (and avoidable) premature failure. They were not entirely incorrect, but they lacked proof.

Airworthiness certification became a nightmare to airlines, because overhaul costs soared, and unexpected failures were often experienced after service work was performed. For example, the Boeing 707, with a 150 passenger capacity, needed 4,000,000 man-hours of overhaul tasks before reaching 20,000 operating hours! Clearly, the viability of commercial airline business economics was now threatened by its own airworthiness certification requirements.

In the 1950s, airlines found that it was not possible to determine an overhaul frequency for the aircraft or more of its components with confidence. Reliability and overhaul frequency were not directly related. Some short-life parts controlled the reliability of some subsystems. They also discovered that eliminating overhaul tasks for parts displaying no clear age-to-reliability relationship decreased maintenance manpower costs without decreasing reliability.

Also in the 1950s, the US Federal Aviation Agency (FAA) was struggling with the fact that some types of aircraft engines continued to fail even with optimized overhaul frequency changes. Then, the FAA started doing statistical analysis on fault events documented by the insurance industry. At that time, the FAA required an approved and well-documented preventive maintenance plan for each airplane type from anyone fabricating and selling aircraft. By the 1960s, Boeing proposed a new plane design with three times the passenger capacity of the largest existing 707 model. In keeping with its policies, the FAA required an acceptable maintenance plan for the 747-100. They argued that existing maintenance programs consisting of time-based rigid inspections, repairs, and replacements would also be three times as expensive and time consuming as compared with the 707 plane model size. The FAA deemed that the model 747-100 aircraft was not economically viable. Boeing and United Airlines (UA) decided to challenge the bathtub wear pattern assumption that was believed to be true throughout the industry of the day.

UA and Boeing then worked on a new process to determine what kind of maintenance assets were needed, given the physics of each failure, to keep airworthiness. A maintenance steering group (MSG) became responsible for developing a systematic common-sense process used to determine what to do to preserve systems functions for the Boeing 747-100 aircraft. MSG developed an acceptable proactive maintenance (PM) program proving both technical and economic viability. It received the endorsement of the FAA. MSG wrote a handbook on the approach used for formulating maintenance strategies for the new Boeing 747. The handbook, published by the US Air Transportation Association (ATA) in 1968, was titled *Maintenance Evaluation and Program Evaluation* (MSG-1). The basis of the handbook was a decision diagram designed for maintenance strategy selection. The document was generalized to be used in other planes in the 1970s. The second document was known as MSG-2 and was titled *Airline/ Manufacturer Maintenance Program Planning Document*. It was used in models Lockheed 1011, Douglas DC-10, and some other military aircraft. In the 1970s, the US Department of Defense (DoD) named the new method *reliability centered maintenance* (RCM). In 1975, DoD required that all major systems be evaluated with RCM.

UA studied its own extensive failure database and proved that there existed more than one failure pattern for non-structural components. Failure density functions of some 230 different non-structural components revealed that the vast majority of such failures were random in nature as demonstrated by patterns d, e, and f of Figure 2.2 and that only 11% of the failures were related to operating age as shown in patterns a, b, and c of Figure 2.2. Moreover, only 4% of the failures obeyed the bathtub curve. Being optimistic, it might be fair to assume that the then current maintenance plans were effective in preventing about 11% of the faults. A pessimistic approach would establish that maintenance plans were right 4% of the time. So, they were off target 96% of the time. This was a landmark finding. UA found six types of failure patterns (Figure 2.2), grouped them into two categories, and published them in their 1979 report "Reliability Centered Maintenance" by Stanley Nowlan (UA's director of maintenance analysis) and Howard Heap (UA's manager of

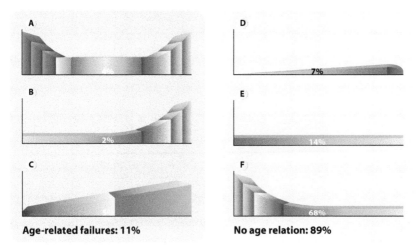

Age-related failures: 11%   No age relation: 89%

FIGURE 2.2
Boeing and UA study result showing six types of failure patterns classified into two groups.

maintenance program planning).[1] In Figure 2.2, the graphs represent conditional probability of failure over time. Pattern A is the bathtub curve previously described. B shows an aging pattern where the likelihood of failure is greater with age. C shows a special case of aging, where resistance to failure decreases steadily, as in erosion, corrosion and fatigue. Pattern D, rapid aging and then random, is found in hydraulic and pneumatic systems. E has a constant failure rate of completely random failures. Finally, F shows the "worst new" situation where system "infant mortality" is a problem, often because of overhauls or other interventions, including new installations.

The UA report was later modified to become MSG-3 (1980) for application of the maintenance process to new Boeing models 757 and 767. A European version of MSG titled *European Maintenance System Guide* was applied to the Concorde and Airbus 330/340 in 1993. The application of RCM principles and feedback provided for air design coming from the analyses resulted in an astounding improvement in safety, from approximately 60 crashes per million takeoffs in the 1960s to just fewer than two in the 1990s (it's 0.2 today). Even more impressive is the fact that the improvement was achieved at a significantly reduced cost. Figure 2.3 illustrates the experience of the aviation industry when it applied MSG-1 to Boeing 747-100 craft as compared with the older DC-8-32.

The nuclear industry started applying RCM. In 1982, the Electric Power Research Institute (EPRI) carried out two pilot projects in nuclear plants.

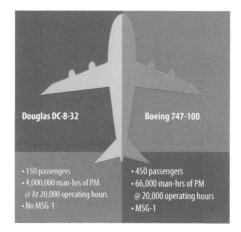

FIGURE 2.3
PM hours requirement comparison before and after MSG-1.

By 1987, it was adopted across the nuclear industry in the United States. By 1993, streamlined versions of the EPRI model were appearing to help with cost savings during the RCM process itself. In 1982, RCM was being applied in mining and other industries in South Africa, and it migrated elsewhere in 1986, first to the United Kingdom, Europe, and North and South America and then to the Middle East and Australasia. The first commercial publications on RCM appeared in 1991[2] and 1993.[3] Following the publication of those books, RCM became increasingly well-known and popular. It was also well suited to the increasing complexity of plants, the growing need for more sophisticated maintenance techniques, increased global competitiveness, and of course, the need for optimizing the output of assets at the lowest possible cost. A confusing array of RCM variations began to appear in the 1990s, some of which were not even close to the level of rigor described by Nowlan and Heap or in later books. In 1999, the Society of Automotive Engineers (SAE) published a new standard (SAE JA-1011, "Evaluation Criteria for Reliability-Centered Maintenance [RCM] Processes") in an effort to clear up the confusion, provide a tool to weed out the less effective methods, and provide an alternative to the then very complex military standards that had emerged. Since then, only a couple of other books have been published on the topic before this one.

Now, let's go back to the question posed at the start of this chapter. An unbiased answer is not difficult to present. Today, the aviation industry is less expensive and much safer than it was six decades ago thanks to the RCM process developed in that industry. That is an undeniable fact. Fortunately, true RCM practitioners (i.e., those following the methodology designed by Nowlan and Heap) also experienced good results when doing RCM analysis on plant and other fleet assets. The authors and their reliability engineering teams have trained many maintenance professionals and have led the application of RCM to hundreds of facilities with a variety of assets in North America, South America, Africa, and Europe. Some of the industries and assets types assessed with RCM include automotive, pharmaceutical, biotechnology, medical devices, water treatment, mining and ore processing, food and beverage, oil and gas production and refining, cement, steel, manufacturing, underwater construction, power generation, transmission and distribution, and transportation, among others. Our experience with RCM is gratifying, as we have seen preventive maintenance time being reduced by 40%–70% for existing programs, while corrective maintenance–related costs have been diminished by as much as 50% when compared with programs that were in place before RCM was applied.

YES, it is probably worthwhile to do RCM at your plant. We will expand the information on RCM benefits in Chapter 12, but not before we discuss how the RCM process can become a futile mission if not applied properly.

RCM is a systematic process to determine what must be done to keep assets doing what operators need them to do in their current operational context. In other words, it is a process of producing effective "failure management policies." It produces maintenance plans by which maintenance tasks are prioritized according to their consequences and targeted specifically at failure causes. It also produces other decisions on operator-performed tasks, one-time changes to procedures or processes, design changes, and even, in the right circumstances, running an asset to failure.

We have to bear in mind that RCM cannot be applied to every asset, and that not all possible causes of failure may need a programmed maintenance task. Thus, RCM requires a complete mindset shift from the traditional maintenance management approach of doing PM to everything everywhere in the hope of preventing all failures. We only apply the necessary maintenance at the lowest cost (and risks) for the asset to do what operations requires of it.

The RCM process requires a multidisciplinary team to answer seven questions about the asset being assessed:

1. *What are the functions and associated desired standards of performance of the asset in its present operating context* (functions)? In other words, what does operations want the asset to do versus what the asset can do at its peak performance level?

2. *In what ways can it fail to fulfill its functions* (functional failures)? We need to define which failures to perform the defined functions are relevant and worthy of in-depth analysis.

3. *What causes each functional failure* (failure modes)? We brainstorm on all possible failure events and their causes.

4. *What happens when each failure occurs* (failure effects)? We must determine each failure's impact by describing the sequence of events happening when each failure mode occurs.

5. *In what way does each failure matter* (failure consequences)? How is safety, environment, production, or maintenance cost impacted? Was the failure a result of a faulty protection device?

6. *What should be done to predict or prevent each failure* (proactive tasks and task intervals)? We must determine whether any form of proactive maintenance, that is, what condition or time base tasks, can be applied to avoid each failure.

7. What should be done if a suitable proactive task cannot be found (default actions)? In the case that proactive tasks are not technically or economically viable, the team has to determine the most appropriate course of action for failure management. At this stage, we must decide whether we can let the failure happen or whether redesign or some other one-time change is needed to avoid the failure or its consequences. Options for these actions will be discussed in Chapters 4 and 11.

RCM is recognized as one of the most powerful tools a company can use to obtain more value from its physical assets. It is the cornerstone of highly successful maintenance programs to ensure that machines help operations to deliver as required, yielding or exceeding the anticipated financial outcome required by stakeholders. It is a means to optimize reliability and maintainability performance to achieve greater uptime. RCM is a process that is also capable of delivering many other benefits as well as just improving reliability.

Enterprises undergoing asset and maintenance management practices transformation will not only see RCM as another project but adopt it as a foundational program to get the most benefit from it well into the future. Improved overall machine knowledge, improved motivation of a better informed and engaged workforce, PM workload reduction, better machine design, enhanced safety, superior environmental awareness and performance, healthier maintenance and operational practices, lower spare parts consumption, and better teamwork (maintenance, production, planning, safety, etc.) are some of the benefits a company embracing RCM can accomplish when it is applied correctly.

Furthermore, it should not be carried out on all assets, but only on those considered critical to the company. But, don't get the impression that RCM can do all this on its own. There may be many other aspects of maintenance and business processes that need to be fixed even before RCM can be applied to your assets and be fully successful. Following RCM, there are also activities that will have to take place to implement the maintenance plans and other recommendations that RCM produces.

Most organizations now realize that maintenance is a far more important aspect of their business than most stakeholders may have imagined. Maybe it was somewhat overlooked when the operation was originally designed and eventual production planning took place. Even today, very few capital projects outside of aircraft, nuclear, and military applications include funding for such a review of what will be required to support operations once commissioning is completed.

We know, as maintenance professionals, that the maintenance cost impact of finished goods can range from 5% to 50% of the total cost incurred to produce them. Today, maintenance is regarded as a valid career path and even as an engineering field. We have bachelor-, master-, and even doctorate-level degrees on the subject. Sometimes, maintenance engineering is combined with reliability engineering to define the new maintenance and reliability engineering field.

These all respond to a growing need for greater knowledge of maintenance, reliability, asset management, life cycle costing, and reliability, availability, maintainability, and systems (RAMS) engineering, just to be able to meet or exceed corporate expectations. Bad maintenance practices are bad news. They can certainly lead companies to tremendous production losses. Also, safety, environmental issues, and even non-conformance to governmental and other important regulations may be faced as the result of poor maintenance and its management.

Is RCM suitable for your organization? This is another key question. Why bother if RCM is not applicable to our organization? Let's look at some relevant matters before we jump into RCM. You can conduct a self-assessment on RCM suitability for your plant by answering the following key questions:

1. Does my organization have a considerable capital investment in machinery, buildings, and vehicles? (Yes or No)
2. Do we need an increase in production and/or a reduction in maintenance cost to stay or be more competitive? (Yes or No)

3. Do we have competitors? (Yes or No)
4. Does a significant production loss result from the failure of just one or a few machines? (Yes or No)
5. Are the most important production units spared or replicated? (Yes, No, Somewhat)
6. Do we have good maintenance and failure events documentation? (Yes, No, Somewhat)
7. Are machine failures seriously affecting production? (Yes, No, Somewhat)
8. Is your organization aware of the need for asset and maintenance management as recommended by the ISO 55000[4] standards and the Uptime[5] methodology practice? (Yes, No, Somewhat)
9. Would upper management support an RCM program at your facilities? (Yes, No, Somewhat)
10. Is your organization ready to adopt a long-term vision on asset and maintenance management? (Yes, No)
11. Are you looking for a long-term solution? (Yes, No)
12. Does your company have trained RCM practitioners, or would it think about certifying some? (Yes, No)

Now, follow the scoring scheme in Table 2.1 to quantify the result of your self-assessment survey.

TABLE 2.1

RCM Viability Self-Assessment

Question Number	Y-Value	SW-Value	N-Value
1	35	*	0
2	5	*	0
3	5	*	0
4	10	*	0
5	0	5	3
6	5	3	0
7	10	5	0
8	5	3	0
9	5	3	0
10	5	*	0
11	5	*	0
12	5	*	0

Then, add the values obtained for the 12 survey questions to obtain the assessment numerical result. If your score equals or exceeds 70, RCM may be very suitable for your organization. A score below 70 but greater than 50 indicates that your organization needs to do some pre-work before starting an RCM effort to get good results afterwards. A result with a score below 50 indicates that RCM is not suitable for your organization. Asset management, maintenance management, RCM, condition-based maintenance root cause failure analysis, spare parts optimization, and key performance indicators monitoring are all needed to be a top performer today.

Not all of this is easy to implement. There are a lot of elements involved in making it work for your organization, many of which touch on other parts of the organization outside of maintenance. Chapter 1 provided an overall idea of asset and maintenance management and just how RCM fits into the whole asset management spectrum. The following chapters describe the RCM process in detail. Also, we explain how true RCM according to the SAE JA1011 standard can be complemented with some other qualitative and quantitative tools to attain even better results. We call our enhanced approach *Reliability Centered Maintenance-Reengineered* (RCM-R®).

REFERENCES

1. F. Stanley Nowlan and Howard F. Heap. *Reliability Centered Maintenance*, December 1978, United Airlines, San Francisco, CA.
2. John Moubray, *Reliability-Centred Maintenance*, 1991, Butterworth-Heinemann Ltd, Oxford, UK.
3. Anthony M. Smith, P.E. *Reliability-Centered Maintenance* 1993, McGraw, New York, USA.
4. International Standards Organization, ISO 55000. *Asset management: Overview, principles and terminology*; ISO 55001. *Asset management: Management systems—Requirements*; ISO 55002. *Asset management: Management systems—Guidelines for the application of ISO 55001*, January 2015, Geneva, Switzerland.
5. John D. Campbell and James Reyes-Picknell. *Uptime: Strategies for Excellence in Maintenance Management*, 3rd edn, 2015, Productivity Press, New York, USA.

3

The RCM-R® Process

THE SAE JA1011 RCM STANDARD

Reliability centered maintenance (RCM) was developed by the commercial aviation industry to improve the reliability, safety, and cost effectiveness of its operations. Stanley Nowlan and Howard Heap, both from United Airlines, documented it in their report published by the US Department of Defense in 1978. The RCM process is based on a common-sense procedure with a decision diagram for creating maintenance strategies to protect assets' functions. Since its origins, RCM has been used in many industries and in almost every industrialized country in the world. There have been many individual interpretations of Nowlan and Heap's report leading to the creation of a variety of methods that differ widely from the original process. The standard SAE JA1011, published in 1999, sets out the criteria that any process must comply with to be called *RCM*.

The standard is based primarily on the RCM process and concepts established in Nowlan and Heap's 1978 report "Reliability-Centered Maintenance." Other documents, such as US naval aviation's MIL-STD-2173, NES 45—Naval Engineering Standard 45, "Requirements for the Application of Reliability-Centered Maintenance Techniques to HM Ships, Royal Fleet Auxiliaries and other Naval Auxiliary Vessels," and John Moubray, "Reliability-Centered Maintenance (RCM 2)," were also used as sources to develop it. The 12-page document, revised in August 2009, describes the minimum criteria for a process to be considered an RCM-compliant method.

The standard provides the criteria to establish whether a given process follows the doctrine of RCM as originally proposed. It can also serve as a guide for organizations seeking RCM training, facilitation, or consulting.

We summarize the step-by-step process of classic RCM,* as described in the SAE JA1011 standard, before explaining the RCM-R method.

Document SAE JA1011, AUG 2009, establishes that for a process to be acknowledged as RCM, it must follow these seven steps in the order shown:

1. Delineate the operational context and the functions and associated desired standards of performance of the asset (operational context and functions)
2. Determine how an asset can fail to fulfill its functions (functional failures)
3. Define the causes of each functional failure (failure modes)
4. Describe what happens when each failure occurs (failure effects)
5. Classify the consequences of failure (failure consequences)
6. Determine what should be performed to predict or prevent each failure (tasks and task intervals)
7. Decide whether other failure management strategies may be more effective (one-time changes)

Operational Context and Functions

The standard is very specific on how to record the functions of the asset under analysis. Bear in mind that the RCM process is common sense driven. Thus, the logical starting point for designing a maintenance or failure management strategy (or an asset management policy as the standard calls it) is understanding clearly what is being demanded from the asset. This represents a change in perspective for maintainers. Often, the maintenance department is not involved in determining why any particular asset is actually there. However, if we are to sustain the performance of specific functions, we need to know exactly what the functions are as well as the operating parameters that define the performance levels needed to fulfill operational demand. Neither author thought much about this when first working as a maintenance engineer.

More than two decades ago, one of us remembers being told by senior maintenance engineers to study the manufacturer's maintenance and operation manuals to decide maintenance tasks for plant assets. Two years later, it became evident that applying similar maintenance plans to

---

* The term classic RCM is often used to describe the RCM process developed by Nowlan and Heap, but has also been used to describe other rigorous RCM methods.

machines of the same kind was not good enough, because some of them still failed. On the other hand, others were in very good shape when over-hauled. Both of us learned, the hard way, that pure time-based overhauls or replacement often induced failures.

That senior engineer was about to retire. He was the utilities department head in charge of operations. Both maintenance and utilities operations were centralized services in that plant. He once said, "I won't let you do this week's PM to that pump." When asked "Why?" with the explanation "Equipment did not have a PM program before I started working here and now the company requires machines to get PM," he smiled and said, "I don't have anything against you or your PM program, young man. The thing is that the last time your group overhauled that pump it broke and it was working ok before the PM." It was very difficult to team up with operations by that time, since there were a lot of barriers between the two departments. Finger pointing at each other was very typical for profes-sionals working in the two different departments. RCM was created some time before its authors clearly established how it had to be carried out. Teamwork is needed for the model to work, since not all the information needed to answer the seven basic questions is known by maintenance alone. To properly define the operating context, the RCM team (main-tenance and operations together) must describe functions following this structure in accordance with the standard:

1. The conditions in which a physical asset or system is anticipated to operate shall be defined, recorded, and available.
2. All primary and secondary functions of the asset/system shall be identified.
3. All function statements shall contain a verb, an object, and a quanti-tative performance standard (whenever possible).
4. The performance standards used in function statements shall be the level of performance desired by the user of the asset in its current operational context. The design capability should not be used in the function statement.

Functional Failures

If you ask a group of persons in a plant to define the term *failure*, they may come up with a variety of answers. The maintainer may very well tell you that he or she considers the asset failed when it is unable to run. The machine operator understands that the machine fails when it is running below design capacity. The production or plant manager could perceive that the process is failing when full production demand is not attained. The SAE JA1011 standard defines functional failure as "a state in which a physical asset or system is unable to perform a specific function to a desired level of performance." Thus, it is instrumental to have a perfect understanding of the asset functions and the desired performance level to determine functional failures. The companion standard to JA-1011 is SAE JA-1012 "A Guide to the Reliability-Centered Maintenance (RCM) Standard." It was published in 2002 to further clarify how to meet the requirements of JA-1011.

There may be total or partial functional failures. This means that the asset may not able to fulfill a particular function at all or that it may perform it at a performance level lower than desired. All three of the answers in the preceding paragraph could be right, considering that each is thinking of different functions and standards of performance. It may not be immediately evident here, but the causes of those three different failures may also be different. The SAE standard asks that all the failed states associated with each function be identified, so that we are able to identify all the relevant failure causes.

Failure Modes

The term *failure mode* is not heard as frequently as *failure*, even among maintenance people. A failure mode is a single event, which causes a functional failure to occur, and each failure mode usually has one or more causes. So, we need to brainstorm on all possible events impairing the ability of assets to perform each specific function to the desired level of performance. This sounds like a lot of work to do. The standards provide some hints to avoid excessive work. Also, they recommend not being too superficial about the causation level of the failure modes. When listing failure modes, consider the following:

1. All failure modes reasonably likely to cause each functional failure shall be identified.
2. The method used to decide what constitutes a "reasonably likely to occur" failure mode shall be acceptable to the owner or user of the asset. Usually, consensus is used to decide which failure modes to analyze and which ones to discard.

3. The level of causation for failure modes must be exhaustive enough that appropriate failure management policies can be assigned to manage them.
4. Failure modes listed in the analysis must consider events that have happened before, the failure modes being prevented in the existing PM program, and other events that are likely to occur in the actual operating context but have never happened.
5. Human and design errors causing failure events must be included in the failure mode list unless they are being addressed by other analysis methods.

Failure Effects

Failure effects quantify the "damage" each particular failure event may cause to the plant or the organization. It is recommended to describe "what happens when the failure mode occurs." The standard recommends several relevant considerations to help understand how serious each particular failure cause might be. We have already defined how the asset works, how it fails, and what exactly caused the failure. In this step, we determine the extent to which each failure mode is relevant by taking into consideration the following:

1. Is there any evidence that the failure has occurred?
2. What is the potential impact the failure poses on personnel safety?
3. What is the potential impact the failure poses on the environment?
4. How is production or the operations affected?
5. Is there any physical damage caused by the failure?
6. Is there anything that must be done to restore the function of the system after the failure?

Failure Consequences

Failure effects are classified into categories based on evidence of failure, impact on safety, the environment, operational capability, and cost. We should be able to decide which of the four categories applies to the effects of each failure mode. Only one category must be chosen—whichever is most severe. Hidden and evident failure modes must be clearly separated. Failures with safety or environmental impact must be distinguished from those having only economic impact by either operational or nonoperational consequences. As with every step within the RCM process, failure consequence determination is critical. Maintenance strategies are carefully selected for every critical failure cause based on a decisional procedure using the failure consequence as the starting point.

Maintenance Strategies Selection

The most likely predominant failure pattern for each identified failure should be taken into account at the time of recommending any failure management strategy. Based on the two groups and six dominant failure patterns shown in Figure 2.2, failure modes may occur with age or usage or randomly. They may also occur prematurely or following a wear-out pattern after some significant operating time. Care must be taken to recommend maintenance tasks based on actual predominant failure patterns. SAE JA1011 acknowledges five possible maintenance strategies that must be applied to mitigate the consequences of any given failure:

1. *Condition-based maintenance tasks*: These tasks are intended for detecting potential failures. Such detection must occur early enough so that corrective action can be taken before the loss of function. A condition monitoring task is applied at fixed intervals to enable trending of the function loss prior to a functional failure. Figure 3.1 shows the PF interval as the time elapsing between the potential and functional failure events. Note that the shortest time to react to a functional failure equals the PF interval minus the task interval. It is identified by the "minimum reaction time" zone in Figure 3.1.

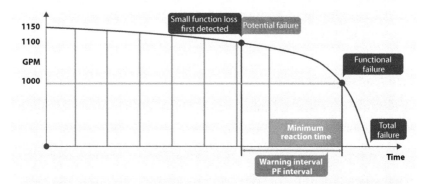

FIGURE 3.1
PF curve.

Hence, we must ensure that the time is long enough to plan and execute a corrective action.

2. *Scheduled overhaul tasks*: Time-based repair tasks must be carried out based on the useful life of the component: that is, the time at which the component failure rate ceases to be constant. Theoretically, at the end of the useful life, the component failure rate increases beyond a rate that we can tolerate. This corresponds to the right ends of the curves A, B, and C of Figure 2.2. Besides the useful life of the item, the cost of the preventive repair also needs to be evaluated. That is, a comparison of the cost of the overhaul work with that of the functional failure must confirm the economic viability of the task.

3. *Scheduled replacement tasks*: Scheduled discard and replacement tasks are considered when it is demonstrated that replacing is more cost effective than overhauling the item. It is recommended to apply such replacement at the end of the so-called "economic" life of the item.

4. *Failure finding tasks*: These tasks are intended to detect hidden failures associated most of the time with protective devices or redundant components. We must ensure that it is physically possible to perform the recommended failure finding task and that the suggested task frequency is acceptable to the owner of the asset. More will be said about task frequency in Chapter 11 and Appendix A.

5. *Redesign tasks*: Sometimes, appropriate time, condition, or failure finding tasks for a critical failure mode can't be found. Then, it may be imperative that modifications (also called *one-time changes*) are implemented to properly address the failure consequences. Changes in assets' physical configuration, operation or maintenance procedures, operator/maintainer training, and operating context alteration are all possible forms of one-time change or redesign potentially required for mitigation of failure consequences. If there are hidden failures with the potential for causing multiple failures having safety or environmental consequences that cannot be detected using failure finding tasks, they must be addressed with redesign tasks capable of reducing the likelihood of the multiple failures to a level that is tolerable to the user. On the other hand, only economic viability evaluation is needed for hidden failures not having safety or environmental impact. If the failure is evident, poses safety or environmental consequences, and cannot be dealt with using proactive maintenance strategies, then a redesign task capable of reducing the

risk to an acceptable level to the user will also be required, whereas evident failures having only economic impact (production loss or maintenance/repair-related costs) are optional and require only that redesign tasks be evaluated with regard to economic viability.

When formulating maintenance tasks, appropriate frequencies must be assigned for them to effectively address failure effects. Some mathematical and statistical formulas are used to support the task interval decision. In such cases, the SAE JA1011 standard recommends that the math used be agreeable to the item's owner. Also, care must be taken when recommending new maintenance tasks for assets, since the RCM process cannot, by any means, supersede existing laws, regulations, and/or contractual obligations without agreement of the governing body imposing those requirements. Thus, it is wise to have a knowledgeable internal auditor evaluate and accept recommendations made as part of the RCM process.

RELIABILITY CENTERED MAINTENANCE-REENGINEERED (RCM-R®)

Both authors have early experience with RCM. In the early 1980s, James worked in a petrochemical plant, where he learned that the best information about how to handle failures generally came from the maintainers, who knew how the assets were actually used. He learned a great deal more about preventive and predictive maintenance and why it was done, something that he had managed but hadn't fully understood in his earlier years as a ship engineer. Later, he would realize that this is a common shortfall—there is little to no education on proactive maintenance explaining just "why" we do it, only "how". Deeper insight and potentially greater ability to get value from it, usually comes from experience. More practically focused education would be needed. In the mid-1980s, he was formally exposed to RCM in a defense project, where he used it to define the maintenance and support program for a fleet of ships. Having been exposed to the complicated military styles of RCM, he gained a healthy respect for the efficiency of using a functional approach as defined by Nowlan and Heap, and Moubray. In 1995, James got into consulting. He has used RCM extensively for the rest of his career and spent a good deal of time learning directly from John Moubray and from working in and with his "Aladon" network of RCM2 practitioners. In 2004, he formed his own consulting firm and has worked with RCM as a part of his "Uptime" model of excellence ever since.

His experience with RCM was that even with good analysis, there were problems in follow-up implementation of the decisions. That was similar to the experience he was having with his "Uptime" model. As soon as the consultants left, efforts trailed off and sometimes stopped. Moreover, as companies were becoming lean and more interested in short-term results, they were less inclined to invest heavily in training and long RCM projects. He began redesigning his approach to delivering RCM and other consulting services and was in the process of doing this when he met Jesús. By then, Jesús had been successfully applying a variation of RCM that he had developed during postgraduate thesis work, and he was about to write a book. James had already produced two of the three editions of his (coauthored) book, and the two agreed to write this book on RCM-R together.

So, how did Jesús get to this point?

By May 1993, Jesús was undergoing a work interview for an "equipment reliability engineer" position at a major chemical plant. He heard the term *reliability centered maintenance* for the first time during the interview, when his future boss was explaining the success RCM had had in the aviation industry. He was coming from a major pharmaceutical plant, where he had helped to implement a predictive maintenance program based on vibration analysis, infrared imaging, oil analysis, and ultrasound monitoring. He learned a great deal about preventive, predictive, and precision maintenance before doing his first RCM analysis. During those years, he also had the chance of managing the maintenance spare parts storeroom together with planning and scheduling. He needed that background to understand later how those pieces are properly put together by means of the RCM process. He always worked within the maintenance department, supporting operations and utilities. Predictive and precision maintenance were his passion, and he loved performing all types of interesting analyses. Another thing he prized was "plant shutdowns." They were often a mess for others, because they required long work hours for some weeks. But for Jesús, shutdowns always represented the best time for learning and applying precision tools during and after repairs.

In 2001, Jesús decided to leave private industry to form his own service company. His company did precision and predictive maintenance at first. It was essentially the same predictive and precision maintenance work as before, albeit all over the country (Puerto Rico), along with delivery of a lot of public and in-house vibration analysis classes in the Caribbean region. He still worked long hours, spending most of his time working at plants and getting his hands dirty, and doing public training. One day, a friend asked an interesting question: "Have you considered teaching how to do predictive and precision maintenance to some new engineers while you move to an upper level in maintenance services?" This guy was not even a maintenance person, but he imagined there must be some "management" area within maintenance. Then, Jesús remembered his ex-boss's words on RCM. He started searching and ended up learning about postgraduate studies in maintenance and reliability engineering. He was eager to study what reliability really meant and wanted to see it from other than the pure maintenance department perspective.

During postgraduate studies, he had had to carry out research and practical projects on some important maintenance and reliability methods, including RCM. During one of those projects, he visited a company that had experienced RCM engineers to learn more about the process and see some analysis documentation. Surprisingly, one of the engineers showing him an RCM report admitted that the analysis had been carried out some years back, and they still had significant failures in these assets. By that time, Jesús had also had exposure to both qualitative and quantitative reliability tools. He asked for the chance to apply his 1 year graduation project thesis to this situation. Both the university and the company agreed, so he formulated his project hypothesis.

Classic RCM analysis provides its practitioners with a clear understanding of asset functionality, its criticality, and the consequences of its possible failures. It also helps in identifying the maintenance tactic that should be employed to tackle each relevant functional failure root cause. RCM practitioners also enhance their knowledge of the asset under study because of the amount of time and effort dedicated to understanding how it works, how it fails, and which type of maintenance should be employed. RCM could be considered a form of suggest replacing with root cause failure analysis (RCFA) before the fact and can be applied to an asset even at the design stage.

When RCM is applied to assets in operation with a failure history, its analysis is best complemented with failure data analysis. Such failure events will turn the analysis team's attention to the most probable failure modes, which are in fact the ones occurring. If failure events are clearly captured and documented in corrective maintenance work orders, important information leading to an understanding of the nature of each relevant failure can be obtained. This will enable even better choices of appropriate maintenance tactics. Important indicators will let us know each asset's current situation through the use of quantitative measurements such as mean time between failures (MTBF), mean time to repair (MTTR), and inherent availability (Ai). Maintenance cost data is also helpful, and together with failure data analysis, sets out the basis for improvement. RCM-R is comprised of five basic elements:

1. Data integrity: The quality of data found in the work orders documented by maintenance personnel must be top notch. If important failure events are not captured properly, vital reliability tools are useless in the plant. Hour meter reading must be referenced for failures together with the as-found failure causes. Time to repair, spare parts, and manpower used constitute part of the relevant information of a true reliability culture aiming to improve equipment uptime.

2. Classic RCM: This is a process that needs no further justification regarding its use for designing appropriate maintenance tasks to critical assets. It is important that practitioners apply it well—they must stick to a method compliant with SAE JA1011. Also, we need to make sure that the analysis group has in-depth knowledge of maintenance, especially modern maintenance techniques. Good FMEA work has resulted in poor maintenance recommendations because of a lack of knowledge on the part of the RCM team members about condition-based maintenance. Often, this is revealed in the form of many time-based restoration and replacement tasks because they can't properly identify condition-based maintenance approaches that might in fact be more suitable.

3. Failure data analysis: Weibull analysis is applied whenever well-documented data is available. Chapter 11 contains more detail on Weibull and other data analysis tools. Such analysis enables the team to find out actual failure patterns instead of having to guess them. Sometimes, we have found that the team is assuming a wear-out pattern when in reality, an item exhibits random failures. The opposite case may also be possible. We've even seen teams assuming premature failures are occurring when in fact, wear-out patterns have been observed. Opinions are valuable, but good numbers don't mislead.

4. RAM analysis: Acceptable uptime and availability are the result of a combination of good reliability and maintainability. So, good work order data analyses will enable us to determine quantitative reliability, maintainability, and availability parameters.

5. Continuous improvement: Change is the only constant. You hear people saying this all the time, and they are right. Processes, asset

demand, and maintenance methods, among many other things, may all change. Also, machine modifications and upgrades can occur. Unexpected failures could also happen all of a sudden. We need to realize that RCM never ends and should become a living process. Failure to do this is at the heart of most classical RCM program failures—great RCM project work goes "stale" over time and becomes less effective if it is not kept ever-fresh through a living process.

RCM-R® PROJECT

Many companies devote significant time and effort to qualitative RCM analysis, leading them to design a comprehensive maintenance program for their maintainable assets. Some enterprises realize later that despite being as robust as it is, RCM is not always able to avoid some component failures impacting significantly on critical assets' Ai. If relevant repair information is properly documented, the most recurrent failure modes can be easily identified. Failure data can be extracted from the historical maintenance records and analyzed, and suitable maintenance actions can be recommended according to their reliability characteristics. All failure types (wear-in, random, wear-out, and combinations of these) can often be discerned if the data is reasonably accurate. Failure management tasks can reduce the failure rates to a minimum in most cases, and the potential increase in machine availability can be determined. Let's get a bit technical now.

Here is a list of symbols used in the mathematics:

Ai = inherent availability
β = shape parameter (Weibull distribution)
η = characteristic life (Weibull distribution)
CBM = condition-based maintenance
CM = corrective maintenance
e = 2.718281828 (the base of the natural logarithms)
F = failure
MTBF = mean time between failures
MTTR = mean time to repair
PM = preventive maintenance
R = reliability
t = time (either time to failure or replacement time)

System Description

Pharmaceutical process air handling and dehumidification units provide a continuous flow of air at a specified volume of flow, specific temperature, relative humidity, and purity. The subsystems of a typical unit are

1. Precooling system
2. Dehumidifier system
3. Cooling coil system/supply air fan
4. Process room
5. Exhaust fan

The functional block diagram in Figure 3.2 shows the interactions among the subsystems and presents some relevant control and protection devices and flows. The acronyms and abbreviations used are

- AC: alternating current (electricity supply)
- CFM: cubic feet per minute of air flow
- Comp Air: compressed air
- PCV: pressure control valve
- PSV: pressure safety valve (relieves excess pressure)

Most of the volume of treated air (14,050 CFM) circulates in a loop with an additional 2,500 CFM of new air entering the loop while that same amount is exhausted.

Reliability, Availability, and Maintainability (RAM) Analysis

Quantitative parameters used to measure reliability maintainability, and availability are defined as follows:

- Availability[1] "is the period of scheduled time for which an asset is capable of performing its specified function." It expresses the probability than an item, under the combined influence of its reliability, maintainability, and maintenance support, will be able to fulfill its required function over a stated period of time and when called on to do so. An asset need not be running to be available, as long as it is capable of running.

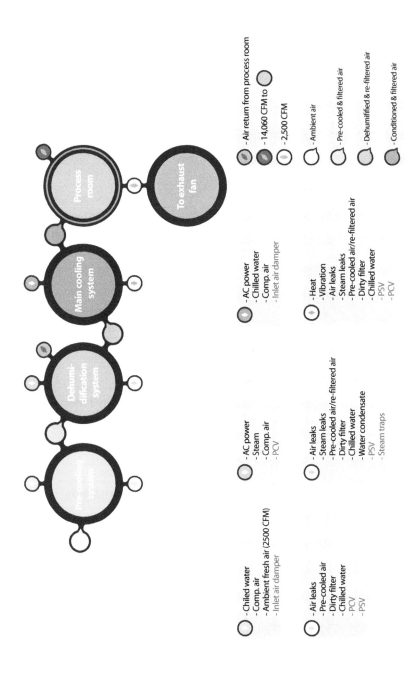

FIGURE 3.2
Air handling and dehumidification system functional block diagram.

$$A = \frac{MTBF}{MTBF + MTTR} \qquad (3.1)$$

- Maintainability[2] is "a measure of the ability to make equipment available after it has failed, or mean time to repair (MTTR)." It is determined by

$$MTTR = \frac{\text{total downtime from failures}}{\text{number of failures}} \qquad (3.2)$$

- Reliability[3] is "a measure of the frequency of downtime, or mean time between failures (MTBF)." It is determined by

$$MTBF = \frac{\text{total operating time}}{\text{number of failures}} \qquad (3.3)$$

Current Situation Explained

Classic RCM analysis was done for the three air handling units (AHUs), resulting in a comprehensive preventive maintenance (PM) program that has been in place since 2006. It included maintenance tasks for every component and instrument identified in the system's technical drawings.

Maintainers check every bolt, valve, filter, rotating component, electrical fuse, and so on. The program is administered through two main semi-annual PM work orders for mechanical and electrical tasks, respectively, and one for a quarterly lubrication PM. Also, vibration spectral analyses are done for all motors and fans monthly to pinpoint incipient rotating component problems.

As part of a broad infrared thermography program, a survey is performed every six months on electrical substations, motor control center cabinets, and programmable logic controller (PLC) boards. An average of 75 PM work hours are dedicated to each unit every year.

There are three AHU and dehumidification systems at the plant, identified as systems A, B, and C. A brief review of work order data led to determination of each system's Ai for the year 2009. See Table 3.1 for details.

PM and corrective maintenance (CM) costs for the year 2009 for each unit are shown in Table 3.2. The two tables are completely related, since both number of failures and MTTR affect the total yearly corrective maintenance cost. This data enables us to see the system's current reliability

TABLE 3.1

System Availability for Year 2009

System ID	# of Failures	MTBF (Days)	MTTR (Hours)	Ai
A	20	18.25	5.27	0.988
B	26	14.04	2.86	0.922
C	18	20.28	3.2	0.993
	64			

TABLE 3.2

Maintenance Cost Data for Year 2009

System ID	PM/CM Time Ratio	Failure Cost per Hour	PM Cost per Hour	Total Failure Costs per Year	PM Program Cost per Year	Total Maintenance Cost per Year
A	0.230	$2,000.00	$200.00	$210,800.00	$4,848.40	$215,648.40
B	0.980	$2,000.00	$200.00	$148,720.00	$14,574.56	$163,294.56
C	0.740	$2,000.00	$200.00	$115,200.00	$8,524.80	$123,724.00
			Totals	$474,720.00	$27,947.76	$502,667.76

performance and the costs associated with achieving it. PM and CM cost data was provided by the plant.

Failure Data Analysis at a Glance

Failure data obtained from the computerized maintenance management system (CMMS) is sorted by failing components in Figure 3.3 and ordered by numbers of failure occurrences, roughly equating to cost-saving

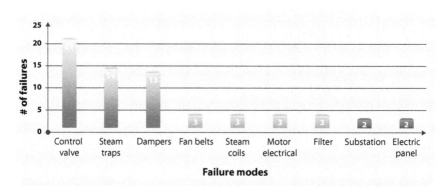

FIGURE 3.3

Components failure data for year 2009.

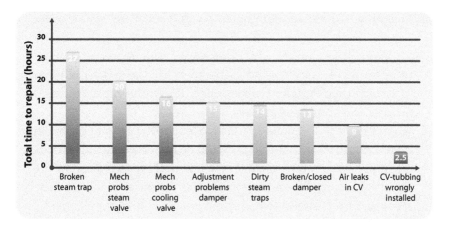

FIGURE 3.4
Most repeated failure modes for year 2009.

opportunities to the organization, if controlled. Figure 3.4 shows the failure data broken into failure modes and ordered by time to repair.

Introduction to Weibull Distribution and Analysis

The primary advantage of Weibull analysis is the ability to provide reasonably accurate analysis and failure forecasts with extremely small samples. The two defining parameters of the Weibull line are the shape parameter, beta (β), and the characteristic life, eta (η).[4] Beta is related to the physics of the failure, and eta is the typical time to failure in Weibull analysis. The two-parameter Weibull distribution is by far the most widely used distribution for life data analysis.[5] The Weibull cumulative distribution function (CDF) provides the probability of failure, $F(t)$, up to time (t):

$$F(t) = 1 - e^{-\left(\frac{t-\gamma}{\eta}\right)^{\beta}} \tag{3.4}$$

$$R(t) = e^{-\left(\frac{t}{\eta}\right)^{\beta}} \tag{3.5}$$

where:
$F(t)$ is the probability of failure up to time (t)
$R(t)$ is reliability, that is, the probability that failure will not occur up to
time (t)

The meaning of the shape parameter[6] β can be interpreted as follows:

β < 1: Premature failures are experienced and repair at failure is recommended. Also, it is advisable to verify the quality of repairs, or an improved design may be needed. Preventive replacements will increase the number of failures.

β = approximately 1: Random failures occur with constant probability. If failures are frequent, design should be improved. Regular replacements are not recommended, as they do nothing to reduce the probability of failure. Monitoring to predict the onset of failure may be appropriate if cost effective.

β = 1 to 3: Wear-out-type failures that show some random failure characteristics. Fixed time replacement not usually recommended. Monitoring is appropriate, and some maintenance at failures may be necessary if cost effective.

β > 3: Wear-out. Fixed time replacement may be effective depending on the cost of the PM as compared with that of the failure.

Based on the Weibull analysis, some components are to be discarded or replaced after some specified operating time. When that is the case, optimum replacement time can be calculated. Two requirements must be met for the preventive replacement of a component to be appropriate. First, PM makes sense when the component condition gets worse with time. In other words, as the component ages, it becomes more susceptible to failure or is subject to wear-out. In reliability terms, this means that the component has an increasing failure rate. The second requirement is that the cost of the preventive replacement must be less than the cost of CM when failure occurs. If both of these requirements are met, then PM is appropriate, and an optimum time (incurring minimum cost) at which the preventive replacement should take place can be computed.[7]

Optimum Replacement Time Analysis

Figure 3.5 shows the concept of total maintenance costs as a sum of corrective and preventive costs. As a rule, whenever preventive activities are reduced, then corrective activities (repair of failed devices) can be expected to increase. The total is usually a curve having a minimum or optimum point.

The optimum replacement or preventive maintenance time is calculated by the use of Equation 3.6:

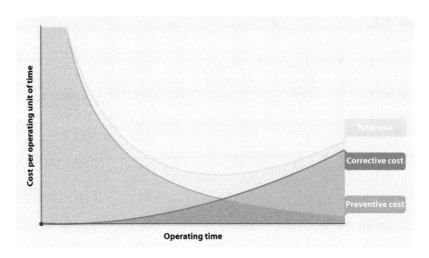

FIGURE 3.5
Computing the optimum replacement time.

$$CPUT(t) = \frac{\text{total expected replacement cost per cycle}}{\text{expected cycle length}}$$

$$CPUT(t) = \frac{Cp * R(t) + Cu * (1 - R(t))}{\int_0^t R(t)\,dt} \qquad (3.6)$$

where:

Cp is the preventive replacement cost
Cu is the corrective replacement cost
R(t) is the probability that failure will not occur up to time (t)

Detailed Failure Data Analysis Using the Weibull Distribution

The most recurrent failure modes have been analyzed using the Weibull distribution. Weibull reports are shown in Table 3.3. An example of a Weibull plot is presented in Figure 3.6.

Table 3.3 summarizes the Weibull parameters for the most frequently recurring failure modes experienced during 2009. Recommended maintenance strategies based on the failure analysis versus the current maintenance strategy are presented in Table 3.4. Optimal replacement times were calculated with the use of RelCode* software as shown in Figure 3.7.

---

* Note that RelCode software has been discontinued by its owners.

TABLE 3.3

Weibull Parameters Summary

Failure Models	Total Time to Repair	Shape Parameter Beta	Characteristic Life (days) Eta	Failure Type
Broken steam trap	27	0.75	86.7	Premature
Mech probs steam valve	20	4.01	508.9	Wear Out
Mech probs cooling valve	16	2.25	195.5	Rand + Wear Out
Adjustment problems damper	15	1.53	195.8	Rand + Wear Out
Dirty steam traps	14	0.96	29.9	Random
Broken/closed damper	13	3.93	169.39	Wear Out

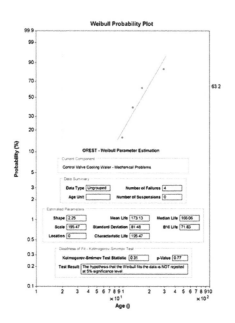

FIGURE 3.6

Cooling water control valve failure.

TABLE 3.4

Summary of Maintenance Strategies Recommendations

Failure Models	Current Maintenance Strategy	Optimal Replacement Age and Unit Cost	Recommended Maintenance Strategy	Task Detail and Frequency	Task Time	% Failing w Policy
Broken steam trap	Insp. PM-180 d	Non	Detect and Redesign	Ck repair quality & Train Techs. ultrasound–90 days	0.5	50
Mech probs steam valve	Insp. PM-180 d	223 days @ $4.78/day	TBM	Replace @ 180 days @ $5.06/day	3	2
Mech probs cooling valve	Insp. PM-180 d	67 days @ $21.76/day	CBM	Valve opening time monitoring and trending–30 days	0.5	1
Adjustment problems damper	Insp. PM-180 d	72 days @ $16.20/day	TBM	Re-adjust @ 90 days @ $16.26/days	1	26
Dirty steam traps	Insp. PM-180 d	Non	Detect and Redesign	Check correct valve application/ Installation design	?	?
Broken/closed damper	Insp. PM-180 d	74 days @ $7.33/day	TBM	Overhaul@90d @ 7.80/days	3	8

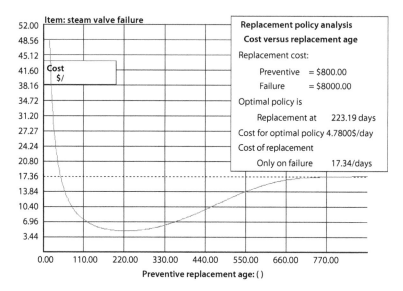

FIGURE 3.7
Steam control valve replacement analysis.

The Weibull and optimal replacement analyses are now combined and enable us to determine and predict maintenance program costs based on the baseline failure data we had for year 2009. Table 3.5 shows the impact that the recommended actions will have on reliability, maintainability, and availability. Note that some 38 failures related to the most common failure modes are avoided with the new policy. There has also been a significant increase in PM-associated costs, but the downtime cost associated with failures is expected to decline by more than half, as shown in Table 3.6. In this case, the most expensive failure modes were targeted for reduction by the new maintenance strategy.

TABLE 3.5

Predicted Systems Availability with New Maintenance Policy

System ID	# of Failures	MTBF (Days)	MTTR (Hours)	Ai
A	12	30.42	2.92	0.9960
B	11	33.18	3.31	0.9959
C	13	28.08	2.77	0.9959
	26			

TABLE 3.6

Predicted Maintenance Cost with New Maintenance Policy

System ID	PM/CM Time Ratio	Failure Cost per Hour	PM Cost per Hour	Total Failure Costs per Year	PM Program Cost per Year	Total Maintenance Cost per Year
A	2.855	$2,000.00	$200.00	$70,080.00	$20,007.84	$90,087.84
B	2.910	$2,000.00	$200.00	$72,820.00	$21,190.62	$94,010.62
C	2.900	$2,000.00	$200.00	$72,020.00	$20,885.80	$92,905.80
			Totals	$214,920.00	$62,084.26	$277,004.26

CONCLUSIONS

Failure data analysis, through the use of the Weibull distribution, enhances the capability of a classic RCM analysis when applied to equipment with failure history, provided the failure events are well documented, such that failure modes and repair times can be identified in the data. The analysis will reveal the physics of the failure, enabling the choice of an appropriate maintenance tactic to address the failure mode and reduce its failure rate. Also, when the CM and PM costs are known, the optimum replacement or PM interval can be determined and implemented to reduce the total maintenance costs associated with a single failure mode.

The results of our specific case for the air handling and dehumidification units revealed that the implementation of the proposed maintenance strategies would yield a potential 59.4% reduction in total failures for the three units, resulting in a total maintenance cost reduction of $215,663.50 annually. The expected equipment availability was determined by subtracting the number of repair hours that would have been avoided in 2009 if the recommended policy had been in place.

THE RCM-R® PROCESS DIAGRAM

It took some time for the outcome of this project to be analyzed and for all the work done to address the customer situation to be organized into a logical process, as shown in Figure 3.8. This particular reliability analysis was performed for a customer with a particular need. But, what if the

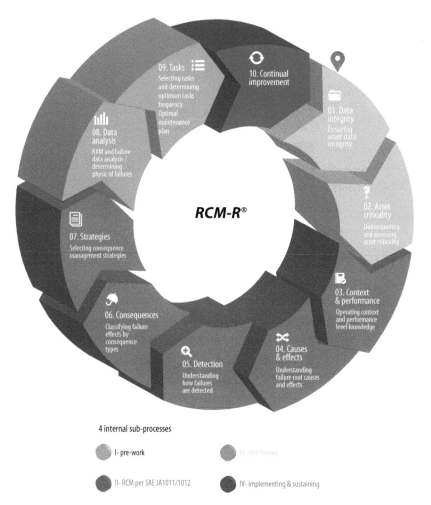

FIGURE 3.8
The 10-step RCM-R® process.

"particular" need of this customer repeated itself in other companies? RCM-R® was developed as a result, as a process to complement classic RCM in accordance with the Society of Automotive Engineers (SAE) JA1011 Standard with quantitative reliability tools. RCM-R is a particularly effective tool when applied to critical assets with a history of failure events. Though accurate failure data is necessary to perform Weibull analyses, the method can be adapted to less precise data if that is all that is available. Interviewing techniques can be used in some cases to elicit sufficient information to make reasonably accurate estimates of Weibull parameters, which can be correlated with the available data. In all cases, however, asset owners are encouraged to improve the quality of failure data documentation for better future reliability analyses.

Each of the ten RCM-R process phases will be explained in detail in the following chapters. Chapter 4 focuses on the aspects to be considered prior to performing an RCM-R exercise on a critical asset. Phases 1 and 2, dealing with asset data integrity and criticality, among other important considerations, will be discussed in that chapter. Chapter 5 concentrates on functions and functional failures, for which the aspects of Phase 3 are of vital importance. Failure types and classes are discussed in that chapter. Chapter 6 deals with part of Phase 4 by discussing failure modes and causes, while the rest of that phase is discussed as part of Chapter 7, dealing with failure effects. Chapter 7 also contains a discussion of Phase 5, leading to understanding how failures are detected. Failure effects are classified by consequence type (Phase 6), as discussed in Chapter 8. Various proactive maintenance techniques are discussed in detail in Chapter 9. Maintenance strategy selection (Phase 7) and the RCM-R decision diagram are discussed in Chapter 10. RCM fine tuning (Chapter 11) focuses on the use of complementary quantitative tools such as RAM and Weibull Analysis (Phase 8) and on determination of task frequencies (Phase 9). Phase 10 is covered in Chapter 12. Now that we have presented the RCM-R methodology, we are ready to explain in detail this whole reengineered process.

REFERENCES

1. John D. Campbell and James V. Reyes-Picknell. *UPTIME: Strategies for Excellence in Maintenance Management*, 3rd edn, 2015, Productivity Press, New York, 456.
2. Emile W. J. Eerens, *Basic Reliability Engineering*, 1st edn, 2003, Le Clochard, Mount Eliza, 22.
3. Campbell and Reyes-Picknell, *UPTIME*, 464.
4. Robert B. Abernethy, *The New Weibull Handbook*, 5th edn, April 2010, Abernethy, Author and Publisher, North Palm Beach, Florida, 1–3.
5. Abernethy, *Weibull*, 2–2 to 2–3.
6. Raymond Beebe, *Machine Condition Monitoring*, 2001 edn, MCM Consultants Pty, 6–7.
7. Reliasoft, *Quarter 2 2000: Issue 1, Optimum Preventive Maintenance Replacement Time for a Single Component*, www.reliasoft.com/newsletter/2Q2000/preventive.htm

4

RCM-R® Pre-work

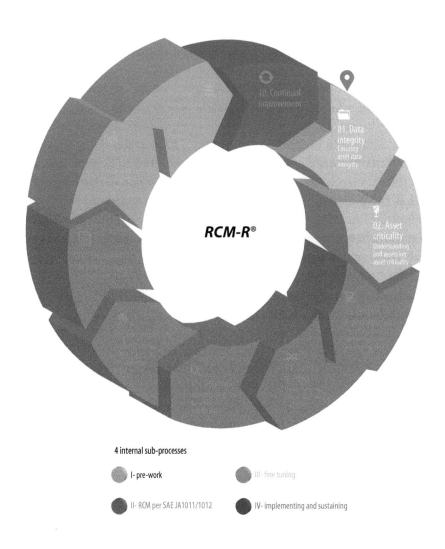

RCM-R®

01. Data integrity
Ensuring asset data integrity

02. Asset criticality
Understanding and assessing asset criticality

10. Continual improvement

4 internal sub-processes

- I- pre-work
- II- RCM per SAE JA1011/1012
- III- fine tuning
- IV- implementing and sustaining

ENSURING ASSET DATA INTEGRITY

Data on its own is more or less useless. Put it into context, and it has meaning. In our context, asset data integrity is all about ensuring that data is accurate and, in the right context, meaningful. Asset data is the collection of facts (data) about a plant asset that provides relevant information to those who require it in a form that is intact, complete, and reliable. What kind of data is the most relevant for proper asset maintenance optimization? We are assuming our RCM-R® analyses are founded on business goals based on stakeholders' expectations as they were conveyed down to the operations level. Maintenance management decisions need to be based on facts converted into relevant information.

Many companies today are data rich (overloaded), yet many of those companies are information poor. The data may not be in its correct context; it may be inaccurate, incomplete, or even missing. There is an abundance of data being accumulated in today's businesses, quite a lot of it related to maintenance and operations, and we need to be very selective about the data we use for our analysis. The value we can derive from asset data has the potential to deliver a significant contribution to your business bottom-line. However, if your asset data is not reliable, your organization could be taking on all kinds of risks and you wouldn't even know it. Merely trusting that data is in good condition is risky, and converting data into information you can trust can be quite complicated.

In an article[1] in *Information Week* (January 2007), writer Marianne Kolbasuk McGee reported the results of a study conducted by Accenture, which surveyed over 1000 managers from US- and UK-based companies with annual revenues of more than $500 million. Their study found that on average, middle managers spend about two hours a day looking for data they need. There are two main reasons for assuming that most of that time is wasted because they do not find what they are looking for on many occasions. First, the volume of data is too large, and most of it is not needed. Second, the quality of data, or its integrity, is generally poor. Much of the data is inaccurate, out of date, inconsistent, incomplete, poorly formatted, or subject to interpretation.

The importance of asset data information is critical to all levels if the business wants to obtain maximum value from its assets. Raw data from a piece of equipment has no direct business value until it is converted into information that is immediately important to the execution of work or making decisions. Good data is like evidence in a courtroom—it is essential to discerning and proving what is happening to inform good decision-making. Thus, success in maintenance and operations depends heavily on proper asset data and information. With it, data analysis can be done, and the data is processed and converted into valuable information. Good information at the workforce level is of vital importance for attaining business goals.

Management requires accurate, usable, and fit-for-purpose information every day to conduct analyses. It is also management's responsibility to prepare concise reports that enable senior management, all the way up to the CEO level, to make sound business decisions. Such decisions, if based on the analysis of reliable data, can render maximum profit for the business. The choices we take at all business levels may also be affected by data provided from external sources. Assets designers, suppliers, and installers need to provide correct and reliable data related to their work, services provided, or equipment supplied. But, at the same time, end users of the assets have to provide precise information on their intended use of asset functionality and its corresponding working environment. Even customer data is very relevant, because it is the whole process driver. The process that should be followed is shown in Figure 4.1.

Let's assume we are examining recommendations coming from an RCM analysis on the most critical rotating machine in the plant. We find that one of the possible failure modes to be addressed is bearing wear due to age. The team was able to reasonably estimate bearing life. Also, the proper task (vibration analysis) was assigned to the asset, and the optimum task frequency was calculated by an "almost infallible" mathematical model. The new maintenance plan was implemented, and predictive maintenance technicians were able to detect incipient bearing wear by analyzing the vibration data taken at the bearing caps. Then, technicians were able to trend the fault over time. Maintenance and production management decided to plan the bearing replacement at a time with no production impact at all. So far, we have applied the asset data process diagram of Figure 4.1 correctly. Thus, if the bearing repair is done properly, the organization will have obtained value by the process of converting raw vibration data into useful information for adequate management decision-making and avoided the consequences of having the critical machine fail unexpectedly and at an inopportune time. Now, let's look at something else that happened in this same incident. When technicians went to the storeroom to get replacement bearings, they found that the

FIGURE 4.1
Asset data process diagram.

bearings stored under the asset register number in the maintenance store-room did not fit the machine. It was found that supplier bearing data was wrong, and the lead time to get the specialized bearings to the plant was 5 days. Repair would have to wait, increasing the risk of machine failure and likely incurring otherwise avoidable plant downtime. What went wrong? An external supplier provided the incorrect information, violating the very first step of the asset data model depicted in Figure 4.1. Even a well-designed-constructed-maintained-operated plant could be at serious risk if incorrect data is provided, processed, and/or analyzed by external sources. In the book *Asset Data Integrity Is Serious Business*[2], the authors explain that asset data has various components. Those components are of vital importance for getting information to help managerial decision-making at all asset life cycle stages. According to DiStefano[2], the different ways data can be classified are

1. *Physical Data*: This type of data describes the type of asset we are referring to. For example, physical data can describe that something is an agitator, motor, heat exchanger, reactor, pump, or any other asset within the plant facilities.

2. *Dimensional Data*: This kind of data is associated with dimensional characteristics and gives the user information on size, shape, weight, etc.

3. *Technical Data:* This sort of data provides the user with more specific information needed for maintenance. For example, material of construction, piping and instrumentation diagrams, and design drawings are all considered technical data.

4. *Reliability Focused Data:* Figures coming from predictive techniques' measurements and information drawn from preventive maintenance work fall within this data category.

5. *Maintenance-Related Data:* Data on repairs focusing on time and material requirements is classified as maintenance related.

6. *Failure Data:* This is data on what actually happened when the equipment failed. Life data analysis is a vital component of RCM-R®, since its outcome showcases the physics of the failure. Such analysis enables management to redesign maintenance tactics accordingly.

7. *Material Data:* Material of construction data is important to operations, maintenance, and environmental health and safety (EHS). Wetted parts materials in pumps are critical for processes handling environmentally dangerous fluids. Therefore, both metallurgical and elastomeric components' characteristics need to be properly chosen to avoid not only production losses but also safety and environmental dangers.

8. *Location Data:* The physical location describes the place within the facilities where the asset resides.

9. *Photographical Data:* Pictures of the asset taken over time can be very valuable for those users who are not able to reach the asset location.

10. *Financial Data:* Expense information on assets can be recorded by way of asset components' financial codes. Maintenance-, operation-, and spare parts consumption-related data falls within this category.

11. *Hierarchical Data:* It is wise to have data on specific components, subsystems, and complete systems. Thus, it is important to understand how the asset is located and related to other components and

systems both physically and from a process standpoint within the facility.

12. *Safety Data:* This kind of data is very important to ensure a safe execution of work on the assets.

13. *Environmental Data:* It is important to analyze possible environmental risks during the operation and maintenance of the asset taking into account past performance.

What is considered good data? The most important aspect is that we feel confidence that the data is correct. If the data we acquired is not correct, neither will be the decision-making resulting from analyzing it. Accuracy is another characteristic that good data possesses: that is, the ability to obtain the correct data the vast majority of times it is collected. Figure 4.2 summarizes the elements of good data.

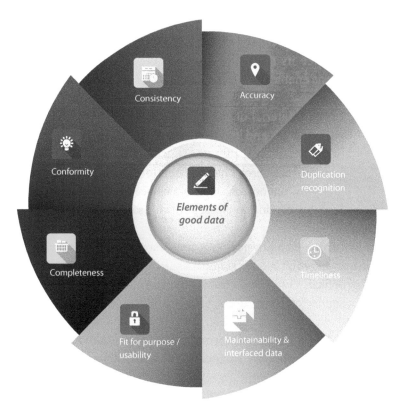

FIGURE 4.2
Characteristics associated with good data.

RCM-R® is a very effective process for determining maintenance tasks for critical physical assets, particularly in the maintenance and operational phase of their life cycle. Thus, data integrity at the task level is a must for attaining optimum asset reliability as a result of an RCM-R® assessment. Relevant data coming from corrective and proactive work orders is a key factor for establishing the current state of a critical asset performance to set out the basis for improvement. Asset data integrity is a complex process existing at both the task and the strategic business levels. A proactive data management process is needed for attaining optimal value from it. If the data management process at a plant is reactive and yields poor data acquisition, less than optimal results will be derived from any effort aimed at improving asset reliability.

RCM-R® DATA

So far, we have mentioned that successful asset and maintenance management relies on analysis and decision-making based on good data. RCM-R® analysis, being a maintenance and asset management tool, also needs specific and reliable data throughout its ten phases to render optimum value to the organization implementing it. Thus, so that we know where we stand with respect to data integrity and whether or not we can trust the data we have, the very first step in RCM-R® is auditing maintenance (and sometimes operations) data collection practices.

RCM-R® combines both qualitative and quantitative analysis for which good data is required. Some phases of the process require very precise data for analysis, while others may just involve the opinions of subject matter experts based on their experience with the asset under analysis. The data needed for each phase of RCM-R® will be explained in further detail in Chapters 7 and 11. The process of converting RCM-R® data into useful information, rendering value to the end user, will also be explained in detail as we describe each step in the whole process. In general, the RCM-R® process requires operational, technical, reliability, maintenance-related, failure, material, financial, safety, and environmental data, which is analyzed for decision-making purposes. The product of the decision-making process is an optimized failure management plan yielding maximum value to the organization. Thus, both maintenance- and operation-related documentation on downtime, spare parts consumption, total PM man-hours, people skills, corrective maintenance man-hours, failure events, quality defects, and so on is needed. System piping and instrumentation diagrams (P&IDs), maintenance and operations manuals, and safety and environmental issues for the assets under study are often needed for conducting successful RCM-R® analyses. A clear understanding of the asset's current state is needed to establish the corresponding desired goals for each RCM-R® project. Post-analysis audits will reveal the level of success the RCM-R® project attained for each particular asset.

ASSET DATA REGISTERS

One of the authors was invited to visit a small manufacturing plant in Europe to discuss a possible implementation of RCM-R® at its facilities. During the visit, he noticed the maintenance manager being interrupted several times while they were speaking. Mechanics were entering his office to get work orders that were being issued in a word processor program and printed by their boss. Each time, the maintenance manager opened a logbook and entered a work order number that he assigned by hand to a piece of equipment that did not have an official asset number while reminding the mechanics to list the parts they would use to repair a failed pump. These observations revealed that they needed to do a lot of pre-work before RCM-R® could be implemented at their facilities. There are some very basic tools that must be in place before any formal maintenance management effort for improving asset reliability can be successfully implemented. A coherent asset data register classification system, often called *taxonomy*, with its corresponding unique asset identifiers is one of those tools needed for the development of an asset management plan and making fact-based maintenance decisions.

An asset data register may range from simple asset lists to detailed information on assets in the form of a technical database with geographic information system links, technical specifications, drawings, video clips on maintenance and repair, and so on. The level of detail of asset registers may vary from company to company. Most firms place a lot of emphasis on financial data, relegating engineering and maintenance information to a lower level of importance, but this will only lead to errors in or excessive effort at engineering and maintenance decision-making. Consider that financial information only records what has been spent, while maintenance and engineering information informs future decisions and actions and then spending. It is arguably more important for keeping the business operations viable. Figure 4.3 illustrates typical information recorded in an asset data register.

Notice that all data components mentioned in the section "Ensuring Asset Data Integrity" are included in the data register system. But, they are organized in different boxes according to the data group they belong to. For example, reliability and maintenance data falls within the computerized maintenance management system (CMMS) data group, while dimensional and technical data is included in the design/project engineering group. Often, these groups of data are actually managed in separate information technology (IT) systems/databases. Having access to all of them and having them consistent with each other is an important consideration.

Data must be recorded in an orderly fashion within its database for better information handling. Storing, recalling, sorting, and analyzing data become more effective when databases are structured following a logical method; otherwise, that structure must be reinvented for every decision or type of decision each time a decision is needed. There are various ways of structuring asset data in hierarchies ranging from three to five levels all the way to a complex nine levels of asset data structure. Standard ISO 14224 "Petroleum, petrochemical and natural gas industries—Collection and exchange of reliability and maintenance data for equipment" recommends a detailed nine-level taxonomy system for oil and gas facilities. Figure 4.4 shows the ISO 14244 recommended asset taxonomy system. A simpler five-level structure provides a practical approach that is recommended for less complex facilities. Figure 4.5 shows a five-level structure taxonomy model as widely used in general manufacturing enterprises.

The following example shows how the ISO 142244 asset data register organizes the information throughout the nine levels of the complete asset hierarchy:

1. Industry = Petrochemical
2. Business category = Petrochemical
3. Installation = Petrochemical complex
4. Plant/Unit = Methanol plant
5. Section/System = Cooling water
6. Equipment unit = Pump
7. Sub-unit = Seal lubrication system
8. Component/Maintainable item = Lubrication oil pump
9. Part = Gasket

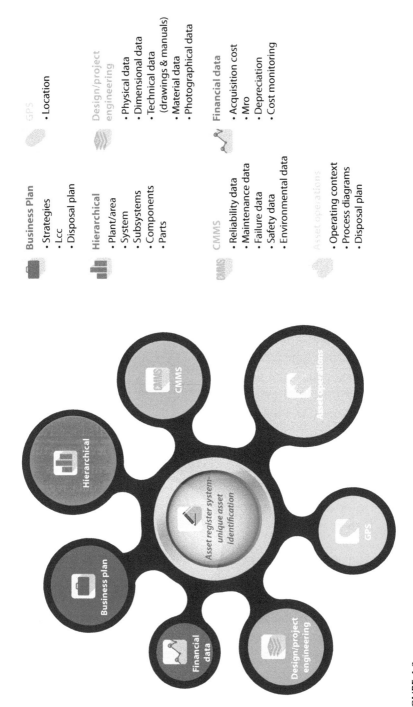

GPS
- Location

Design/project engineering
- Physical data
- Dimensional data
- Technical data (drawings & manuals)
- Material data
- Photographical data

Financial data
- Acquisition cost
- Mro
- Depreciation
- Cost monitoring

Business Plan
- Strategies
- Lcc
- Disposal plan

Hierarchical
- Plant/area
- System
- Subsystems
- Components
- Parts

CMMS
- Reliability data
- Maintenance data
- Failure data
- Safety data
- Environmental data

Asset operations
- Operating context
- Process diagrams
- Disposal plan

FIGURE 4.3
Asset data register information.

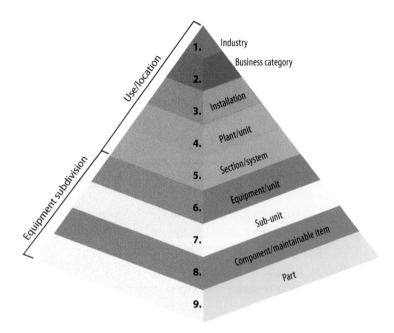

FIGURE 4.4
Asset data register information according to ISO 14224.

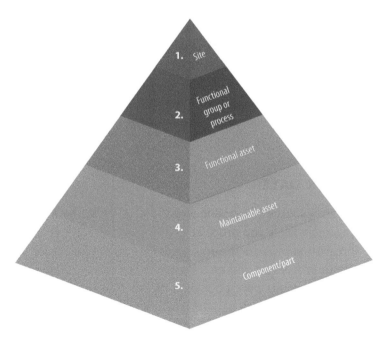

FIGURE 4.5
Asset data register information = five level.

The asset data register system shown in Figure 4.5 starts at the plant or geographical level (1), followed by the productive unit or functional asset group (2). Then, assets or machines with specific functions (3) are identified, followed by maintainable items (4). Finally, parts assemblies or single components, where typically maintenance actions take place, are presented. See the following example, for which the simple five-level structure is used.

1. Site = Houston 3
2. Process = Methylene chloride distillation
3. Functional asset = Reflux pump
4. Maintainable asset = Induction motor
5. Component = Oil seal

Asset numbers or unique identifiers are tied to the type of asset register used by the organization regardless of the complexity of the asset register system chosen. Typically, single items that are not repairable, such as bearings, fuses, seals, small couplings, and cartridge filters, are not assigned an asset number but a maintenance part stock item code instead. Thus, the pump motor mentioned in the asset data register system example may end up with only a four-level equipment number or asset identifier. Therefore, H3-MCD-P1-M would be a possible asset number identifying the motor of Pump number 1 that serves the CH_2Cl_2 distillation column located at the plant named Houston 3. There must be a lot of information related to this P1 unit, ranging from design to maintenance, and all of this information may be recorded under its unique asset number.

WORK ORDER DATA

Work orders are a source of valuable data for RCM-R® and other maintenance management tools and processes. Therefore, the same rule regarding data integrity applies to the information drawn from work orders for reliability improvement and analysis of critical assets. Unfortunately, most maintenance plants have poor data recorded in their maintenance work orders. Comments such as "the unit was repaired" or "the motor was checked" are often found when auditing maintenance work orders for critical assets. The information obtained from such data is so vague that it is impossible for management to make good decisions based on it.

For instance, what failure led to the need for the repair that was carried out, or was there even a failed component found? We just can't tell from the vague work order entry. Ensuring good, useful data does increase the maintenance program effort in the hope that the symptoms reported by operations (e.g., unit isn't working properly) will be resolved and possibly even avoided by future PM tasks. The premise of the RCM-R® process is that failure data documented in corrective work orders can be statistically analyzed to find the predominant failure pattern of each critical failure. Then, the RCM analysis can be fine-tuned with statistical failure data analysis for better maintenance strategies and tasks interval assignment.

It is a necessary part of the process that failure causes of the failing component are identified and the running time is known precisely. If this is the case, effective data analysis leading to proper decision-making regarding maintenance strategies can take place. Work order data is the framework of any good reliability improvement program. Therefore, maintenance and reliability engineers must make sure that important failure and repair data is included in their critical assets corrective and proactive work orders. Standard ISO 14224 recommends recording this data on critical assets for maintenance purposes, and there is no better place to include it than the maintenance work orders:

1. Equipment identification/location: Include subunits and maintainable items intervened (being maintained)
2. Failure code (not relevant for preventive maintenance)
3. Date when maintenance action was undertaken or planned (start date)
4. Maintenance category (corrective, preventive)
5. Maintenance priority (high, medium, or low priority)
6. Description of maintenance activity including failure cause and main findings
7. Maintenance impact on plant operations (zero, partial, or total)
8. Maintenance data spare part consumption (location and availability)
9. Maintenance man-hours per discipline (mechanical, electrical, instrument, others)
10. Maintenance man-hours, total maintenance man-hours, maintenance resources used
11. Time duration for active maintenance work being done on the equipment
12. Maintenance times, maintenance delays/problems

The RCM-R® process converts good data coming out of work orders into useful information, allowing the RCM-R® project team to make the best maintenance task choices for the asset under study.

ASSET CRITICALITY ANALYSIS

An asset criticality analysis (ACA) is an assessment tool to evaluate how asset failures may impact organization objectives. When a business decides to implement an asset management program and has a coherent asset data register system, then ACA is a recommended next step for prioritizing assets for RCM-R® analysis. Also, ACA provides focus to ensure that reliability improvements are made based on calculated risks rather than individual perceptions. Though there is no international standard for asset criticality ranking, there are standards for failure risk management.

- International Standards ISO 14224 (Petroleum, petrochemical and natural gas industries: Collection and exchange of reliability and maintenance data for equipment)
- ISO 31000 (Risk management: principles and guidelines)
- IEC 60812 (Analysis and techniques for systems reliability: Procedure for failure mode and effects analysis)

These provide valuable information on failure risk calculation that can be exploited in designing a good asset criticality ranking assessment tool. Consequently, comprehensive standard-based ACA methodologies have been developed by a number of private firms.

Standard ISO 31000 defines risk as the effect of uncertainty on objectives. Objectives can be related to different business activities, such as financial and EHS goals. They can also apply to diverse organizational levels ranging from departmental, to plant, and all the way up to business. Risk links the combination of potential events (failures) and their consequences to an organization. Thus, risk is expressed in terms of a combination of the consequences of an event and the associated likelihood of the event's occurrence.

ACA is also a process for ordering or ranking assets based on the impact their failures have on the organization's goals. Executing ACA encompasses the assembly of a multidisciplinary team from operations, maintenance, engineering, quality, and EHS. Other departments such as design and materials may also be required in some cases. The multi-disciplinary approach is required to ensure an accurate asset criticality ranking analysis overcoming personal perceptions about the criticality of specific assets. The process should be as objective as possible. The team follows a logical step-by-step procedure for the evaluation of each asset under consideration, as depicted in Figure 4.6.

Risk matrix criteria, as shown in Figure 4.7, are informed by both consequences and likelihood constituents. The four basic components of asset failures' consequences for the ACA are operation, maintenance, safety (and health), and environment related. The team evaluates how asset failures affect the organization's objectives as regards the impact of these four aspects combined. These four components may be subdivided into many more to create a more complex consequence matrix. For example, operations on its own can be subdivided into operational throughput, use, and so on. Also, maintenance consequences can include reliability aspects and can be subdivided into mean time between failures (MTBF), mean time to repair (MTTR), mean cost to repair (MCTR), detectability, and spare parts aspects (availability, cost, lead time, shelf life, etc.). There are many possible combinations when creating a risk matrix for ACA purposes. The most important aspect, however, is that the team ultimately agrees on one model, and that the system owner feels confident it works well for his or her assets. Figure 4.7 shows a generic ACA matrix. Some ACA matrices include multipliers (numerical weighing factors) for each type of consequence. For example, you may use 1 for maintenance consequences, 2 for production-related, 3 for environmental, and 4 for safety consequences, respectively in Figure 4.9. Once again, the RCM-R® team must agree on the criteria used for ranking analysis, and the criticality matrix is an important part of this.

It is important to bear in mind that risk analysis is normally performed on failure events rather than assets. Therefore, it is vital to establish a state of failure considering a reasonably likely worst-case scenario in which the asset completely loses its functionality. It is also worthwhile to carry out the analysis at the highest hierarchy level first and then come down to the maintainable asset if needed. Therefore, ACA for a single plant must be executed at the functional group or process level first. It may be the case that a specific system owner requires an ACA for his or her area later on.

Let's assume the utilities manager of a manufacturing plant wants to carry out the asset criticality ranking exercise for his or her area of responsibility only. The ACA results will be used for prioritizing assets

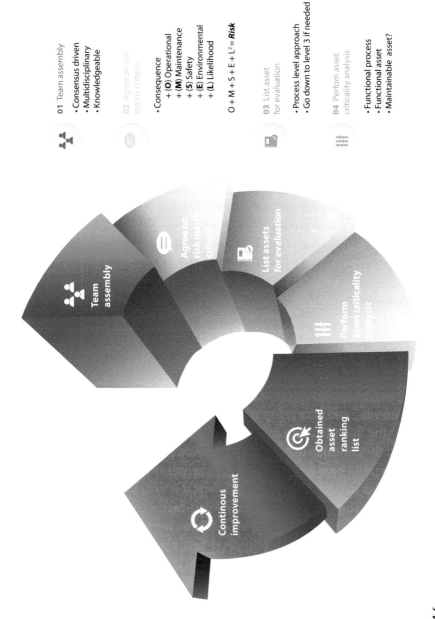

01 Team assembly
- Consensus driven
- Multidisciplinary
- Knowledgeable

02 Agree on risk matrix criteria
- Consequence
  + (**O**) Operational
  + (**M**) Maintenance
  + (**S**) Safety
  + (**E**) Environmental
  + (**L**) Likelihood

O + M + S + E + L² = **_Risk_**

03 List asset for evaluation
- Process level approach
- Go down to level 3 if needed

04 Perfom asset criticality analysis
- Functional process
- Functional asset
- Maintainable asset?

Team assembly

Agree on risk matrix criteria

List assets for evaluation

Perform asset criticality analysis

Obtained asset ranking list

Continous improvement

FIGURE 4.6

Generic asset criticality analysis process diagram.

Simple asset criticality analysis matrix

		Consequence types				
Value	Severity level	Safety	Environmental	Production	Maintenance	Likelihood
0	No impact	No effect	No pollution	No stop	No cost	> 5 years
1	Minor	*Injuries not requiring medical treatment * No effect on safety function	Minor pollution	Minor production stop	Low maintenance cost	1 year ≤ x < 5 years
2	Moderate	*Injuries requiring medical treatment * Limited effect on safety function	Some pollution	Production stop below acceptable limit	Maintenance cost at or below normal acceptance	1 month ≤ x < 1 year
3	Severe	*Serious personnel injury * Potential for loss of safety function	Significant pollution	Production stop above acceptable limit	Maintenance cost above normal acceptance	1 week ≤ x < 1 month
4	Catastrophic	*Lost of lives *Vital safety-critical systems inoperative	Major pollution	Extensive stop in production	Very high maintenance cost	x < 1 Week

FIGURE 4.7
Generic asset criticality analysis matrix.

for RCM-R® analysis and the application of other maintenance management tools. He or she must first assemble a multidisciplinary team for the assessment of the utilities area subsystems according to Figure 4.6. Then, the team must agree on a risk matrix to be used for the risk assessment. They decide to develop one similar to Figure 4.7. They make some modifications to the consequences descriptions and likelihood criteria to make it a better fit to their particular needs. Figure 4.8 shows a possible ACA risk matrix conforming to their particular needs.

Then, the team make a list of all the utilities subsystems to be analyzed for asset criticality. Next, they perform the criticality analysis by using the risk criteria designed and chosen by the multidisciplinary team. The result of this hypothetical analysis is shown in Figure 4.9. Notice that multipliers were used to place different weights on each of the four consequences types, safety being weighted more than the other three. These weighting factors are multiplied by the values obtained for each consequence type assessment. Then, the sum of the values resulting from each consequence type assessment is added to the square of the likelihood value. The total sum for each asset is recorded under the total score column. Once all assets under study are assessed, they can be ranked relative to each other. The far right column was used to rank the five subsystems from top to bottom in order of priority for RCM-R® analysis.

This formula was derived empirically after studying three simple options: multiplying the sum of weighted consequences by likelihood, adding them all together, and adding the consequences to the square of likelihood. In those comparisons, it was the last that best matched customer perceptions of ranking based on their experience with the assets being examined.

Note that there are four colored blocks at the bottom of the table identifying four possible groups into which assets can be classified. This ranking system can be of further use within the maintenance department. For instance, work order priority can be assigned by taking into consideration the criticality of systems, subsystems, and maintainable items together with the urgency of the need for work. This helps the planning, scheduling, and procurement functions.

CHAPTER SUMMARY

There are some important aspects an organization should consider before implementing any formal maintenance management tool, RCM-R® included. We consider two main subjects having to do with good practices

		Simple asset criticality analysis matrix					
			Consequence types				
Value	Severity level	Safety	Environmental	Production	Maintenance	Likelihood
0	No impact	No effect	No pollution	No Stop	No cost	Failure never heard off
1	Minor	*Injuries not requiring medical treatment *No effect on safety function	Minor pollution	x < 2% of plant capacity	Maintenance cost x < $25,000.00	Failure occurred in industry
2	Moderate	*Injuries requiring medical treatment * Limited effect on safety function	Some pollution	2% < x < 20% of plant capacity	Maintenance cost $25k < x < $50k	Failure known in company
3	Severe	*Serious personnel injury * Potential for loss of safety function	Significant pollution	20% < X < 50% of plant capacity	Maintenance cost $50k < x < $500k	About 1 failure per year in system
4	Catastrophic	*Lost of fives * Vital safety-critical systems inoperative	Major pollution	x < 50% of plant capacity	Maintenance cost x > $500k	Several failures per year in systems

FIGURE 4.8

Asset criticality analysis matrix for utilities plant example.

	Abc manufacturing - asset criticality analysis of utilities subsystems						
	Consequence types						
Subsystem	Safety (X4)	Environmental (X3)	Production (X2)	Maintenance (X1)	Likelihood	Total score	Subsystem ranking
Chilled water	0	1	2	2	3	18	4
Cooling tower water	0	0	2	2	3	15	5
Steam 125 psia	3	1	3	3	3	33	2
Compressed air	2	1	3	3	3	29	3
Power generation	4	2	4	4	2	38	1
Ranking group	Critical		Important		Essential		Noncritical

FIGURE 4.9
Example of a hypothetical asset criticality analysis.

that are essential pre-work needed to ensure RCM-R® success. Asset data integrity is of vital importance, since the outcome of bad data analysis could be catastrophic to any organization. The data processing endeavor encompasses collecting, processing, and analyzing raw data to feed management decision-making. ISO 14224 is one standard that recommends which data to collect for reliability analysis purposes. It also shows us how data registers, used to organize data in a hierarchical fashion, are designed for better data handling throughout all asset life cycle phases. Unique asset identifiers, or simply asset numbers, must be assigned following formal asset register formats. Maintenance work orders are a vital source of valuable information for maintenance management and asset reliability optimization. Thus, data gathered from work orders for the purpose of reliability analysis must fully conform to the characteristics mentioned in Figure 4.2 to be considered good. Finally, ACA, often called simply *criticality ranking*, is identified as an essential exercise prior to commencing any major maintenance management tool implementation. It is used mainly to prioritize assets according to the impact their failure may have on the organization's objectives. Using the rankings, we work on those assets that emerge as most critical to the business. The ACA sets out the order in which RCM-R® or any other high-end optimization methodology should be implemented at the facilities.

REFERENCES

1. Marianne Kolbasuk McGee, Managers have too much information, do too little sharing, says study, January 2007, *Information Week Magazine*, New York, USA. http://www.informationweek.com/managers-have-too-much-information-do-too-little-sharing-says-study/d/d-id/1050334
2. Robert S. DiStefano, *Asset Data Integrity Is Serious Business*, 2011, Industrial, New York, USA.

5

Functions and Failures

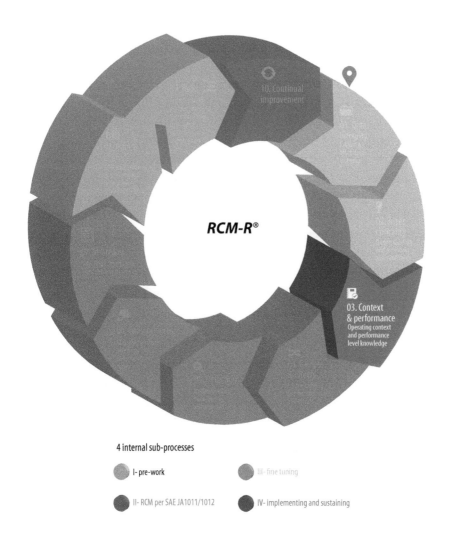

RCM-R®

03. Context
& performance
Operating context
and performance
level knowledge

4 internal sub-processes

● I- pre-work

● II- RCM per SAE JA1011/1012

● III- fine tuning

● IV- implementing and sustaining

THE OPERATING CONTEXT

Sometimes, the authors ask RCM class participants questions on day-to-day aspects of life to help them to grasp technical concepts. For example, we might ask participants to indicate how many tires a Mercedes Benz has, or some similar questions to make them think about the concept of operating context. Most reply that a Mercedes Benz has five tires. It is true that most of the Mercedes Benz cars we see on the highway or at the shopping malls have four tires installed and one more as a backup in the trunk. There are many different but still correct answers to the question, depending on the type of Mercedes we are referring to. Mercedes sedans, motorcycles, vans, trucks, and Formula One cars may require from 2 to more than 20 tires. Cars are often useful examples when speaking about RCM concepts, since most people are familiar with their operation and maintenance. Even similar cars require different maintenance plans depending on their intended duty. In Chapter 2, we defined RCM as a "process to determine what kind of maintenance an asset needs to do what operators require from it in its present operating context." Complete understanding of an asset's operating context is a major requirement for RCM-R® analysis to yield significant value to its end user. This means that the analysis must be performed for the asset in its current application.

We have seen the replication of existing maintenance plans to new plant assets regardless of their application. This is quite a common practice in many facilities today. However, this practice is not compatible with the principle of understanding the operating context prior to assigning maintenance tasks to assets. Let's think about how we apply maintenance to the tires of a sedan car. Sedan cars will have five tires, four of which are running while the other one is kept as a spare in the trunk of the car. What kind of maintenance do we do to them? Almost every car user applies maintenance according to their operating context even without knowing about RCM-R®. Diligent users will normally check their car's running tires for wear, balancing, alignment, and air pressure. Tires actually wear due to friction against the road asphalt. Also, it is generally accepted that if tires are aligned properly, they will wear evenly and last longer. It is also believed that tire pressure should be kept at the manufacturer's recommended values for best performance and lower fuel consumption. But, what kind of maintenance do users apply to the spare tire located in the trunk of the car? Well, even if it is the same kind as the four road tires, its operating context is quite different.

It is not subjected to wear, because it is a backup tire. Its operating context is that of a typical redundant or standby component. The driver does not need to worry about wear and tear; he/she needs to ensure the tire is ready when needed to operate as required. Spare tires can deflate, and if unused for a long time, the rubber gets brittle and less capable of carrying a car safely. Therefore, the maintenance tasks applied in this case ensure that the spare tire is inflated to the recommended air pressure, and that if it gets to six years of age without being used, it is considered for replacement.

The operating context is a statement clearly delineating the environment the assets are intended to operate in as well as an overall description of how and where they are going to be operated. According to the standard SAE JA1011, "the operating context of the asset shall be defined." The operating context should include specifically the type of process (continuous or intermittent) within which the asset will operate. Quality, safety, and environmental standards are other aspects that shall be included in the operating context. RCM-R®, as a process compliant with SAE JAE1011, must consider all these vital facets of the assets when defining their operating context. An ammonia compressor in a beverage bottling plant performing well enough to keep the prime temperature within the required range may not conform to the operating context due to a gas leak. Thus, not achieving production requirements may not necessarily be the only failures considered by RCM-R®. Whether or not the failure is critical needs to be further investigated, and it is the responsibility of the RCM-R® analysis team to find a failure management policy capable of mitigating the potential or imminent risks the failure may represent to the organization. Assets are sometimes subjected to extreme operating environments and locations. The authors have performed RCM-R® analyses on physical assets operating at moderate ambient temperatures of 75°F–80°F at a particular location of a large corporation, and the analysis has yielded a specific maintenance plan. The same analysis carried out at a different location of the same corporation for a similar application but operating at extreme ambient temperatures yielded quite different results as regards the maintenance plan. Extreme environmental temperatures (either cold or hot) may result in different lubricant selection, more accelerated material degradation, greater cooling media flow demand, shaft misalignment caused by thermal growth, and so on. Combustion engines operated at different altitudes burn fuel at different air/fuel proportions, rates, and temperatures to achieve the same engine output. Therefore, the characteristics of the location in which the equipment is to be operated (arctic vs. tropical, desert vs. jungle, onshore vs. offshore, proximity to sources of supply of parts and/or labor, etc.) are of vital importance and must be included in the operating context of assets. Workloads and other operating parameters may vary widely among similar assets within a plant. Is rotating equipment running at full or half speed? Are power generation assets operating under peak load or base load conditions? RCM-R® practitioners must take into account the intensity of operation of the assets under consideration. Some processes are run with plenty of redundancy, whereas others lack any backup at all. Even backup assets may share some support systems (such as the electric power source, for example) with primary units. Thus, backup or standby capability and expected availability are relevant aspects to be included in the operating context. Other considerations, including the need for work-in-process stock to allow repair time without affecting production, the availability of spares for repair, market demands (low or high production seasons), and the availability of raw material supply, are of vital importance when establishing the operating context of an asset.

PERFORMANCE LEVELS

Performance levels (often referred to as the *performance standard*) are another piece of information of vital importance for RCM-R® analysis. The performance standard defines what level of performance the user wants or needs the asset to achieve. It is those standards that RCM-R® will use in considering whether or not any particular function has reached a point of failure. Figure 5.1 shows the concept of performance standards. Note that there is a gap between the desired (what the user wants the asset to do) and

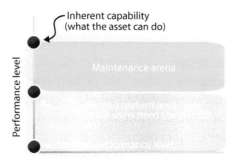

FIGURE 5.1
Desired versus inherent performance levels.

the inherent capability (what the asset could initially do). Maintenance is projected to keep the asset's instantaneous capability within the identified maintenance arena area in Figure 5.1.

The RCM-R® process focuses on understanding the *desired performance level* (what the user wants the asset to do). Maintenance will keep the asset's performance level constantly above, rather than exactly at, the initial performance level or design capability (what the asset can do). If we attempted to maintain initial or design capabilities, then the machine would undergo a lot of over-maintenance, wasting resources and providing less value to the organization.

Thus, the operating context and performance standards are needed to define the asset's functions. Assets operate in a unique manner according to the process they are in and the level of performance demanded from them. The gap between the asset's desired and inherent capability, depicted in Figure 5.1, is called the *margin of deterioration* in SAE JA1012. Design engineers must ensure the margin of deterioration is reasonably large to allow the unit to operate long enough before its components fail to fulfill their functions. The design must also be cost effective. Over-design avoids risking failure of the assets for a long time, but at the expense of high acquisition and operating costs. For obvious reasons, the asset's inherent capability must be above the desired performance level; if not, then the asset is effectively unmaintainable—it will never operate reliably without change. Sadly, this situation is more common than you might expect.

The authors have found recurrent asset failures that are the result of poor design rather than deficient operation or maintenance practices. One of the authors experienced the case of a scrubber pump in a chemical plant failing to achieve the caustic water demand flow and discharge pressure. Gas scrubbers or air purifiers neutralize fugitive emissions coming out of process rooms to clean them before emission into the atmosphere. The author was at home when the shift mechanic called him to report the situation. The author suspected there was a design problem related to the size of the impeller for supplying the required flow, and did some calculations. The result yielded a required impeller size of 7.25 inches in diameter, yet the unit was equipped with a smaller one. Since there were impellers of nominal 8 in size in stock, one of them was machined to 7.50 in to allow a reasonable deterioration margin. In that case, maintenance had been wrongly blamed for the pump's inability to fulfill its primary function, even though it was caused by a design flaw. The scrubber pump was not technically a maintainable asset, because its desired performance level was higher than its design or inherent capabilities as supplied.

Pumps supply a required flow of liquid at sufficient head to overcome downstream resistance to flow as demanded by their process. Some of them operate continually, delivering at defined performance levels (flow rate, pressure, etc.) for some defined time. Others operate intermittently on demand according to some other process parameters (level, temperature, pH, etc.). Maintenance tasks are recommended according to their operating context and performance standard. Assets handling multiple processes have *variable performance levels*. For instance, you may have a pump handling solvents of different specific gravities, pH levels, temperatures, and so on. Its output will vary in each case. The RCM-R® analysis team must decide, together with the asset owner, how to define the operating context and level of performance for that multipurpose asset for the purpose of analysis. A wise approach the authors use often is to consider the worst (or toughest) operating context and performance level case scenario for the analysis. In doing so, the RCM-R® analysis team can warranty the resulting maintenance plan as being capable of protecting the asset's functions well. This is a practical approach if the process demand is quite variable.

It is important to specify *quantitative performance levels* whenever possible. However, sometimes only *qualitative performance* definition is possible. For example, a bench drill may be used to produce holes in parts with a diameter of 0.5000 in ± 0.0010 in of tolerance. It is important for the user that the holes keep their diameters within the defined measurement range, because otherwise they would be rejected by quality control later on for lack of compliance with the manufacturing standards. The acceptable diameter range is a well-defined quantitative performance standard with upper and lower limits on its own. If the user requires an "adequate" surface finishing free of burrs and cutting tool marks, then the operator will just need to ensure that none are left on the parts. But, he or she does not need to warrant that the surface finishing is kept at some defined range of roughness values. This "mark-free" finishing parameter is considered a qualitative performance level in this example.

Some functions may refer to requirements the asset must comply with but which are not possible to describe with either quantitative or qualitative performance level parameters. This "binary" or "go/no go" type of performance level is referred to as an *absolute performance level*. For example, a process pipeline in a chemical process plant consisting of straight pipes, flanges, and gaskets must contain the product flowing inside it. This function (to contain the product inside the piping system) implies zero leakage throughout the entire piping system. The user expects nothing to leak out of the pipe system, meaning that the escape of just a few drops per minute from the system represents a functional failure. If the pump using that piping system is intended to deliver at least 15,000 gallons per production shift, then for that system, only a lower limit performance level has been set. Quantitative performance levels may express the need to perform either over or under some established value for the selected parameter (flow, pressure, temperature, etc.). Some assets must perform between lower and upper limits. Statements such as "to supply between 50 and 60 gallons per minute of water" and "to keep the room temperature within a 70–78°F range" are examples of *lower and upper limit performance levels*.

FUNCTIONAL ANALYSIS

What are the functions and associated performance levels of an asset in its present operating context? This is the starting point of RCM when applied to critical assets. This reminds the author (Jesús) of an emergency service call received from a predictive maintenance services customer. The service engineer asked the customer to explain what situation was found in the unit. The inquiry was made because the fan that was in distress had been included as "OK" on a recent report generated by their vibration monitoring program. The service engineer was told that when the fan was operated at full speed (3550 RPM), it vibrated very roughly. But, the test speed had always been around 1500 RPM during previous periodic vibration data collection. Investigating further, the service engineer learned that the customer tested the machine at a speed way above its normal operation speed, causing resonance in some structural components. There really wasn't a problem at all! The machine operating context was changed temporarily and only for the test. When the unit was returned to normal operating conditions (operating context and performance levels) no problem was found at all. The main or primary function of the blower was to supply 10,000–10,500 CFM of air continuously. That is what the user needed from it, and maintenance only needed to ensure the fan was able to attain this performance level. When operated at full speed, even though the main function had not changed, the increased speed increased the level of output and with it, structural vibration. In this case, different maintenance would be required if the unit were to be operated regularly at full speed. In this case, the unit was run outside the operating range, and there was no need for further analysis. The fan did not need any balancing or structural base redesign.

Assets may have primary, secondary, and hidden (often called protective) functions according to RCM-R®. Maintenance is responsible for ensuring that assets continue to operate at the performance level defined by the owner. It is imperative that the functions, along with their corresponding performance levels described in an RCM-R® analysis, be accurate. Maintenance activities will be selected to protect the assets' functions to mitigate the risks posed by their failures. The RCM-R® process methodology recognizes the importance of functions and is particularly meticulous at describing them for physical assets. Simple assets may have only a few functions, but more complex assets may perform dozens of functions when all primary, secondary, and protective functions are considered. Functional block diagrams are used to better define the functions of large or complex systems by structuring each block's operating context alongside its performance standards. Functional block diagrams are further discussed in the section "Functional Block Diagrams" of this chapter.

Functions must be correctly described during an RCM-R® exercise. A function statement consists of a verb, an object, and a performance level. Whenever possible, performance levels should contain quantitative parameters. A function statement must clearly describe what the asset must do, how it is supposed to be done, and for how long the function needs to be carried out. Functions must clearly establish level of performance as well as efficiency, appearance, quality, safety, and environmental standards, among others, to be met. All function statements shall contain a verb, an object, and a performance standard (quantified in every case where this can be done) as per JA1011 5.1.3. RCM-R® breaks up a complex function statement into multiple statements containing only a single performance level. Subdividing functions as much as possible facilitates failure mode identification by concentrating on a narrower function of the machine each time. The traditional approach to describing functions reduces most primary function analysis to a single function. RCM-R® segments the statement into multiple single performance-level function statements. The best way to organize the functional analysis is to list and number all of the functions (primary, secondary, and hidden) that assets have according to their current operating context.

PRIMARY FUNCTIONS

Primary functions describe the main tasks assets are intended to perform according to their operating context. Primary functions establish the main reasons why a physical asset was acquired by its owner. A pump supplies (or pumps) x flow rate of a certain liquid at y pressure for z time continuously. Air compressors deliver (or compress) x volume rate of air at y sustained pressure for z amount of time. Tanks contain substances in either gas, liquid, or mixed state. These are examples of primary functions for some common assets.

The pump in Figure 5.2 is intended to deliver at least 30 gallons per minute (gpm) at a minimum TDH (total dynamic head) of 100 ft of water, or 35 psi, continuously for 12 consecutive months. If we are applying RCM-R® analysis to this pump, we must first have a good understanding of its operating context before listing its functions as follows:

1. Type of process: continuous.
2. Quality standards: none.

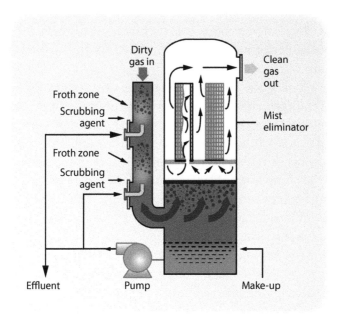

FIGURE 5.2
Gas scrubber water circulation pump.

3. Safety standards: coupling guard, motor cooling fan guard.
4. Environmental standard: zero prime leaks, containment pit, others. Clean Air Act compliant.
5. Location: building X roof with no elevator access (crane needed for overhauls).
6. Asset speed: fixed (full speed).
7. Work-in-process stock: not applicable. A failure shuts down the whole business unit.
8. Asset spares: The asset is fully spared but not backed up with redundancy.
9. Market demand: nonseasonal or uniform.
10. Reliability/availability goals: to operate continuously for 12 months.

Once the RCM-R® team understand the operating context, they can proceed to list all asset functions starting with the primary functions. It is a good idea to try to visualize the process mechanics in slow motion or step by step to perform a detailed functional analysis. Then, the analysis team focuses on shorter but targeted function statements fully compliant with SAE JA1011. This distinctive way of applying the functional analysis will result in a more efficient identification of failure causes later in the analysis. Focusing on more specific function statements relating to just one level of performance optimizes the RCM process and helps the analysis team.

Let's take a look at the step-by-step process the scrubber pump undergoes to deliver its primary function. Before the unit can supply the needed water flow, it has to be able to take it from the make-up reservoir. Thus, the very first primary function to be listed for the purpose of analysis is its suction capability. It is obvious to a pump operator or maintenance engineer that if the pump is not able to pull the water from the make-up tank, it cannot provide the needed flow. But later, this simple function will easily lead us to identify the most probable causes of water not being admitted into the suction line. The next task the pump carries out after admitting water into the suction line is to supply caustic water at the flow and pressure levels the process demands. Since two quantitative parameters are involved in this step, we split the function into two separate functions to consider each of the parameters independently. After the pump delivers water according to the flow rate and pressure demanded by the process, it must keep doing so for 12 months continuously. Therefore, there is a separate function regarding this particular requirement. The complete primary functions statement list will read as follows:

1. To draw caustic water from the make-up tank
2. To supply a minimum of 30 gpm of caustic water
3. To keep the caustic water discharge pressure at a minimum of 35 psia (pounds per square inch absolute)
4. To operate continuously for 12 months

SECONDARY FUNCTIONS

Assets must conform to some safety, environmental, and efficiency requirements while performing their primary function. They are also required to have certain design characteristics offering operating flexibility for performing other than the main or primary function. For example, some manufacturing machines operate at a continuous high speed while in production. But, they must be required to operate in "jog mode" for cleaning between product change-overs. Even ergonomic requirements are demanded from machines, enabling maintainers and operators to execute their tasks with the asset. These other functions assets are required to do are called *secondary functions*. When one of the authors took his first RCM course many years ago, the class trainer recommended the participants to relate the secondary functions to the acronym PEACHES. There are 10 categories of secondary functions, each one related to a letter in the word "peaches," as follows:

- P stands for protection
- E stands for efficiency and economy
- A stands for appearance
- C stands for control, containment, comfort
- H stands for health or safety
- E stands for environmental integrity
- S stands for structural integrity and superfluous functions

RCM-R® facilitators must help the analysis team to identify the secondary functions of the asset under analysis in an orderly fashion. Many experienced facilitators lead the team to brainstorm on each of the seven categories to find the secondary functions for the assets under analysis. Other facilitators consider each component in the system and identify their functions as they step through the system. Generally, most components have only one or two functions. In doing this, there is no need to repeat a function that may have already been described in relation to another component. Later, failures of that component will be dealt with as failure modes. Regardless of the method used, these functions are less obvious than the primary functions, but their loss may have similar or even more serious consequences than a loss of a primary function. Consider the pumping of hazardous fluids—loss of pumping capacity may have economic impacts on a business, while loss of containment (a secondary function) may have serious safety or health consequences.

PROTECTION

Physical assets are often protected from catastrophic failures that could have serious consequences for the plant, the organization, and the community in general. Such catastrophic events are often caused by the loss of protective device functions in conjunction with the loss of whatever function was being protected. Protective devices are designed to protect the safety of people, the product, or the integrity of the asset itself. Safety relief valves protect the integrity of pressure vessels, while proximity probes on safety guards protect the safety of a machine operator by stopping the machine when the guard is opened. It is very common to find many protective and control devices in modern manufacturing assets. Therefore, it is the responsibility of the RCM-R® analysis team to identify each of them and determine their relevance with regard to the integrity of product, people, or assets. Typical protective functions include

- Warning operators: through the use of such devices as audible alarms and flashing lights
- Automatically shutting down a component when it fails
- Eliminating or relieving abnormal conditions after failure: as is the case of overload relays, fuses, rupture disks, and safety relief valves
- Taking over from a function when it fails: by using redundant components

- Preventing dangerous situations from developing in any way—as is the case of machine guards

EFFICIENCY AND ECONOMY

Pneumatic cylinders may still operate with leaks if you increase the air supply line pressure for some time. Many other systems in the plant may be able to work with air leaks, or if their air supply lines pressure are increased or decreased. But, imagine that all pneumatic systems in a plant operated while leaking profusely. In such operating conditions, air compressors would operate loaded all the time, and their electrical consumption would be enormous. Wearing components of the compressors would degrade much sooner, causing more frequent mechanical failures and likely increasing production downtime. Efficiency and economy are closely related when considered in the RCM-R® analysis. Sometimes, high-efficiency motors are required by a corporation as a design standard to keep power consumption to a minimum. Maximum allowed scrap levels can also be defined for processes yielding scrap material. Function statements regarding efficiency standards, such as maximum allowed fuel consumption for stationary engines or even boilers, are often used in RCM-R® analysis. In other words, asset owners expect their assets to fulfill their primary functions in an efficient and economical way, and this is a matter of importance for the RCM-R® process.

APPEARANCE

Appearance is a completely subjective concept driven by the opinion of the asset owner. The maintenance department may like site buildings to be painted gray, but the factory owners may decide that they should be dark brown, because a color consultant determined that brown fits better as a background for the organization's logotype. Thus, company owners believe the image of the company is better projected if they use that color and, of course, if the building's paintwork is kept in acceptable condition.

One of the authors worked with a water treatment facility in an affluent suburb of San Francisco. The facility was designed to fit in with the surrounding architecture and landscape. Not only did they want to keep their presence "low key;" they also wanted to make sure that local residents were comfortable in knowing their water came from such a pristine-looking plant.

CONTROL, CONTAINMENT, AND COMFORT

Sometimes, asset owners or operators need their assets to operate at different performance levels for changed products or processes. They may also need to regulate some operational parameters, such as speeds, pressures, flow rates, levels, and so on. Therefore, the owner expects the asset to *control* such parameters within a desired range for achieving certain special operational tasks. Processes dealing with fluids of any type, such as filling machines in a pharmaceutical or bottling plant, for example, may be required to maintain the prime inside the system pipes and valves even with intermittent operation.

Solvent tanks, pumps, and pipework are also required to operate free of leaks. *Containment* functions are of vital importance for the asset's owners in many circumstances, especially when the fluid is expensive, or when leaks may cause severe consequences to the ambient environment or the company image.

Ergonomics (or human factors) is the scientific discipline concerned with the understanding of interactions among humans and other elements of a system, including physical assets such as machines. Human factors and ergonomics are concerned with the "fit" between the user, the equipment, and their environments. It takes account of the user's capabilities and limitations in seeking to ensure that tasks, functions, information, and the environment suit each user. To assess the fit between a person and the technology, human factors specialists or ergonomists consider the job being done and the demands on the user, the equipment used, its size, its shape, and how appropriate it is for the task. Poor work area visibility, uncomfortable machine seats, unpleasant room temperatures, and restricted working areas cause discomfort to workers. Employee morale and motivation are also affected when such working conditions prevail. Some undesired consequences may be experienced if assets and their surroundings fail to offer an adequate level of *comfort* to their users and maintainers. Occupational accidents and even reliability and maintainability loss are some of the consequences associated with lack of comfort.

Ergonomics also deals with matters of maintainability. Assets must be capable of being maintained efficiently, or they will become economic burdens whenever they require work or repair. Not only must they be operable; they must be accessible and capable of being disassembled or moved to where they can be maintained without major disruption to surrounding equipment and systems. If maintenance or operating adjustments must be made, the doors, guards, switches, handles, levers, and so on that are moved or used in making adjustments must be accessible and workable by human hands. It is not uncommon to find designs, often produced by designers having little field experience, that violate some of these concepts.

HEALTH AND SAFETY

The purpose of RCM-R® is to eliminate failures or reduce their consequences to a tolerable level. In other words, RCM-R® deals with risk mitigation through the use of a systematic process to find failure (or risk) management strategies. Risk, according to standard ISO 31000: 2009, is defined as the effect of uncertainty on objectives, which can be related to different aspects such as financial, health and safety, and environmental goals. Successful businesses always place extraordinary emphasis on having robust *health and safety* standards, practices, and goals. The general manager of one plant asked management to shut down the plant at any time issues threatening workers' health and safety were found. "We simply stop the plant until we feel confident the safety related risk is properly identified, assessed and mitigated," he said. All of the firm's employees felt proud of that plant manager, because he really walked the talk regarding the values the company professed. Users expect assets to fulfill their main functions to attain corporative financial aims but without affecting the health of their users, their maintainers, and the community in general.

ENVIRONMENTAL INTEGRITY

Asset management also encompasses social responsibility throughout the whole life cycle of assets. Physical assets undergo construction, installation, operation, maintenance, and decommission activities, which on occasions, represent potential environmental risks that need to be managed adequately. Improper management of environmental risks may result in tremendous financial, image, and reputation losses to companies. But, the potential for adverse impact on natural resources is of upmost concern here, as no money can fix it. A mega-corporation can pay millions of dollars in fines for causing environmental damages. But, this can't make up for the damage, as it may take decades for mother nature to recover from such incidents. The case of the underwater oil spill caused by British Petroleum (BP) in the Gulf of Mexico in April 2010 is one example of a disaster that could in all likelihood have been averted and could certainly have been managed better once it occurred. Sadly, there are many more examples of the environment being taken for granted and significantly harmed by poor practices in asset management. *Environmental integrity* is an aspect of major concern for RCM-R® and for any SAE JA1011-compliant method. Thus, the analysis team must identify ways in which assets deal with environmental integrity. As responsible and value-driven users, we may find such functions for our assets to fulfill their main mission (as they say in the military arena) while protecting the environment.

STRUCTURAL INTEGRITY AND SUPERFLUOUS FUNCTIONS

Most physical assets comprise systems embracing rotating and fixed equipment such as motors, pumps, engines, turbines, vessels, piping, electrical apparatus, interconnecting pipes and wiring, controls, and so on. We often forget that these components are somehow fixed, bolted, anchored, or attached to the ground. Furthermore, structural components are rarely considered seriously when initial maintenance plans are assigned to new physical assets. We naturally care for ensuring that bearing wear is detected on time. But, we seldom check the integrity of the rotating machine foundations. A lot of care is given to the bolts' torque and base leveling when rotating machines or tanks are initially installed and commissioned. But, after the asset is started up, maintenance is concentrated on other aspects more closely related to the asset's operation. Tasks such as cleaning (because the machine gets dirty when operated), wear checking (because friction happens while in operation), alignment verification (because it gets lost due to vibration during the operation), lubrication (because lubricants get consumed and degraded during operation), and filter changing (because they get dusty due to operation) are never out of sight of maintenance. But one day, a machine's mounting structure collapses all of the sudden. One author has seen entire conveyor systems in mines collapse while in use, access stairways fall from the side of a building, and foundations crumble due to vibration from machinery mounted on them. What happened?

Things as simple as loose bolts, cracked welds, or corrosion triggered the machine structure to break, causing multiple components to fail, generating tremendous production losses and even occupational accidents to the operation. All too often, maintenance programs were developed from "manufacturers' recommendations." Those recommendations rarely consider operating context, but seldom do we see manuals or instructions pertaining to the structures that support our plants and equipment. In the absence of original instructions, these important elements are often forgotten—until one day they fail. Therefore, *structural integrity*–related functions are always of major concern for an RCM-R® analysis.

Superfluous functions are often related to components having trivial roles. Sometimes, machines are provisionally modified or retrofitted with a particular element to tackle some temporary needs or to accommodate the needs of various different customers. A particular soda bottle capping machine was installed with a little water jet nozzle to remove any product drops from the capped bottle necks at the exit conveyor. The water line had a solenoid valve to let the water run when the capper was operated. The RCM-R® team included the function of this component and later learned that the team declared loss of that function "not critical." RCM-R® does not analyze noncritical failures once they are identified. In that case, the operator was asked during the analysis about what happened if the water jet was not working. He replied: "nothing at all; since we just continue operating the capper because there are no quality control requirements for having this device in place." He went on to explain: "it was installed temporarily because one of the machine's base nuts was missing and we thought the machine vibration could cause some dripping during the capping process." Long ago they replaced the bolt, but the temporary water line did not get removed.

Identifying such functions may seem a waste of time, but consider that maintaining those functions consumes resources and time. In a tissue converting operation in the southern United States, one author worked with a team on a pilot project involving one of 14 kitchen towel machines.

Converted paper products such as kitchen towels are usually very high margin products for paper companies, and they can sell pretty much all they can produce. Each towel machine was covered with a large steel "guard" from end to end. Any time work was done on the machines requiring the use of an overhead crane, the guard had to be removed, laid down, and later replaced. That required several hours of work, during which the machine was down, and no one was allowed to work beneath it. On average, each machine needed such work about once every two weeks. In total, the guards were responsible for a loss of about 24 h of production time every two weeks—7% of production losses. When we asked about the function of the guards, we learned they were there to protect against water condensing on chilled water pipes overhead. Drips into the machines could easily ruin product. When the operator described this, a maintainer on the team spoke up and told us that those pipes had been removed some ten years prior, when the old air conditioning system was replaced with a series of chiller units on the roof. There were no more drips, but no one had thought to remove the guards! Needless to say, this awakening resulted in removal and scrapping of the guards, with a 7% increase in production capacity. It was very much worth our while to identify the functions of those guards. Removal of the superfluous function paid for the project!

HIDDEN FUNCTIONS

A function whose failure on its own does not become evident to the operator(s) under normal circumstances is defined by SAE JA1012 as a *hidden function*. Most protective devices and redundant components have hidden functions. Because they protect from some undesirable consequence when something else fails, their failure can have a tremendous impact on their owners. It is the responsibility of the RCM-R® to identify all of the asset components having this type of function. The system's piping and instrumentation diagrams (P&IDs) often provide most of the information on protective devices that the analysis group need to list. Warning lights, alarms, shutdown mechanisms, relief mechanisms, fire suppression systems, life preserving devices, warning signs, and standby components are examples of components with hidden functions. Evident functional failures become totally obvious to the operators under normal circumstances. RCM-R® clearly separates hidden functions from evident

RCM-R®	Company: XYZ System: Exhaust air Component: Scrubber #5 water pump	Location: Town, country Sub system: Scrubber #5 By: Team members	Building: XYZ Date: dd/mm/yyyy

Function #	Function, performance level and operating context
1	To draw caustic water from the make-up tank
2	To supply a minimum of 30 gpm of caustic water
3	To keep the caustic water discharge pressure at minimum of 35 psia
4	To operate continuously for 12 months
5	To handle maximum capacity demand at less than 75% of FLA (Efficiency)
6	To count with good condition painting for process lines, motor and pump comforming to company PT-ABC standard (Appearance)
7	To contain caustic water inside pump volute, pipes and tank (Containment)
8	To count with safety guards in good condition and well installed as per company standard SF-ABC (Heath & Safety)
9	To contain oil or prime leaks inside pit (Environment)
10	To keep appropiate anchorage measured as main support bolts torque of X lb-ft +/- 10%
11	To shut down, two words the motor if load exceeds 125 amps (Protective)
12	To shut down, two words the pump if caustic water tank level is below 6 in. (Protective)

Primary functions	Secondary functions	Hidden functions

FIGURE 5.3
Gas scrubber water circulation pump functions list.

functions, and they are highlighted in the analysis worksheet. Figure 5.3 shows how functions are properly copied and listed in the RCM-R® worksheet. Notice the use of colored backgrounds, which serve as boundaries for segregating primary from secondary and hidden functions.

FUNCTIONAL BLOCK DIAGRAMS

Functional analyses for complex physical assets are better performed when assisted by functional block diagrams. Functional block diagrams describe the functions and interrelationships between several components within a system. A functional analysis must clearly establish the system boundaries, inputs (raw material, ingredients, etc.), outcomes (product the raw material was converted into), supplies (utilities, fuels, etc.), and wastes

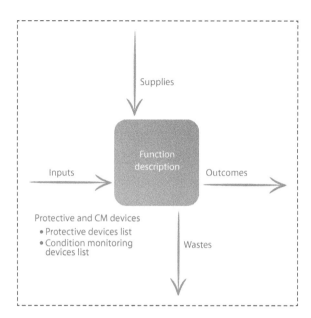

FIGURE 5.4
Simple functional block diagram with boundaries.

(leaks, heat, vibration, etc.). RCM-R® uses a simplified but very practical technique for drawing functional block diagrams, as shown in Figure 5.4. Each block must describe a function performed by a subsystem, including all quantitative (and sometimes qualitative) process parameters. Therefore, inputs and outcome demands are very well depicted by the use of a short statement describing the function inside the block and a group of figures with their corresponding engineering units representing the normal operating conditions as needed by the user. Supplies and wastes are described likewise most of the time.

Once the system block diagram is developed and understood, then the assets' functions can be easily described and listed. The purpose of including the list of protective devices and condition monitoring indicators is to avoid excluding components with hidden functions from the analysis. The mere inclusion of all this vital information in the process functional block diagram triggers the definition of functions, failures, and causes of failures for analysis participants. Functional block diagrams should be meticulously drawn and posted on the analysis room walls throughout the RCM-R® exercise. Figure 5.5 showcases an actual functional block diagram for a pharmaceutical product granulator. Note that the process entails the interaction of five subsystems with several utilities supplies and over 50 protective devices.

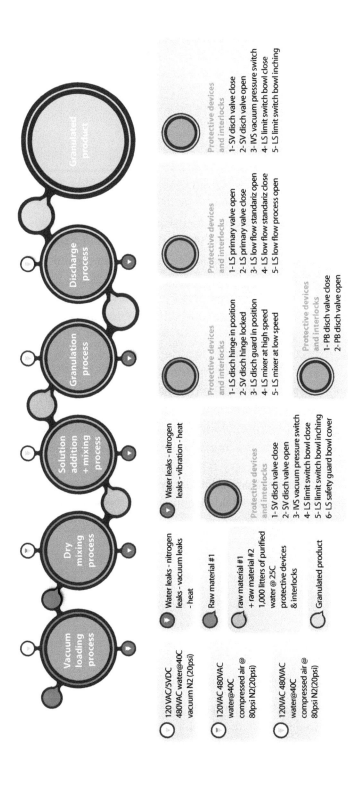

FIGURE 5.5

Actual functional block diagram for a pharmaceutical granulator.

The analysis team used the system P&IDs and the operating manual to put together this functional block diagram. The complete analysis for this system with its corresponding deliverables took several weeks to complete.

FAILURE TYPES AND CLASSES

Assets in general are goods, both tangible and intangible, that represent a potential value to their owners. RCM-R® is most concerned with physical assets and the protection of their functions, allowing owners to obtain maximum value from them. Physical assets are instrumental in attaining or exceeding the organization's financial, safety, and environmental goals. Functional failures occur whenever assets are not able to fulfill their functions at the required performance level. In Chapter 3, we defined functional failure as "a state in which a physical asset or system is unable to perform a specific function to a desired level of performance." Note that the definition of failure mentions the inability to perform "a specific function"—all types of functions are included. Some analyses claiming compliance with SAE JA1011 only care about primary functions. Their focus is on ensuring that the machine delivers what the organization needs from it for increased profitability. Time is almost always the reason for not applying most RCM processes to mitigating the consequences of all critical failures related to every type of function (primary, secondary, and hidden). It is important that maintenance and reliability professionals become very familiar with the field jargon, especially with regard to the variations of the term *failure*.

In RCM-R®, we have two types and three classes of failures. Potential and functional failures are recognized types of failure in RCM-R®. Functional failures may be classified into critical, noncritical, and hidden failures.

TYPES OF FAILURES: FUNCTIONAL AND POTENTIAL FAILURES

Functional failures occur when an asset loses the ability to perform any function to a required performance level. Function #2 for scrubber caustic water pump in Figure 5.3 says: "to supply a minimum of 30 gpm of caustic water." We can establish the functional failure statement by simply negating the functional phrase for every function. This can be accomplished just by adding the word *unable* before functional statements. Then, the expression "unable to supply a minimum of 30 gpm of caustic water" gets our job done. Functional failure statements seem simple, but they are not necessarily straightforward all the time. There are two ways to comply with them. The functional failure statement "unable to supply a minimum of 30 gpm of caustic water" is true if the pump is unable to supply any water at all. It is also true if it supplies less than the required 30 gpm water flow. When an asset loses its ability to perform the function at all, we refer to it as a *total failure*. On the other hand, *partial failure* is what we call a functional failure when the asset is still able to perform at a level lower than the desired performance.

Figure 5.6 shows both cases of functional failure, in which the caustic water pump is unable either to supply water at all or to provide the minimum 30 gpm requirement. Note that a functional failure condition starts just at the time when the asset performance falls below the minimum performance level required by its user.

Some functional failures occur suddenly, giving no signs of functional degradation. Most electronic components, such as fuses, light bulbs, boards, and diodes, do not allow users to detect the gradual degradation of their function. Effectively, they continue to perform at the initial performance level till the time of failure. In this case, potential failures are

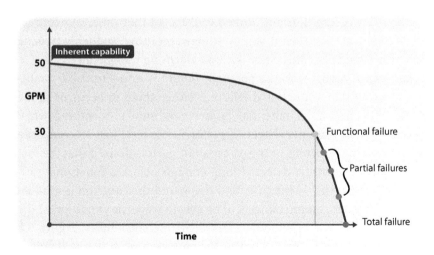

FIGURE 5.6
Functional failures: Both partial and total.

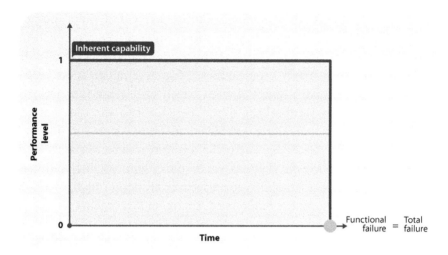

FIGURE 5.7
Purely total functional failures.

never found, and the functional failure is practically a total failure on its own. Figure 5.7 shows the "degradation" curve for a failure showing no signs of performance loss. Let's suppose the third shift operator of our scrubber system found that there was no lighting at all in the roof area when he got up there for a routine inspection at midnight. Investigation by maintenance electricians found that an electronic circuit board for the lighting system had blown, blacking out the area. It happens that the new lighting system was installed six months ago during the yearly maintenance shutdown. The lighting worked well for all that time, showing no signs of deterioration; then it simply stopped working, suddenly passing from a functional to a failed state without alarming the operator.

According to SAE JA-1011, a *potential failure* is an identifiable condition indicating that a functional failure is either about to occur or in the process of occurring. A potential failure condition is identified when an operator or maintainer detects a performance level loss as compared with the system's inherent or initial capabilities. It is desired that the loss of performance event be detected long enough before a functional failure takes place. If that is the case, and the loss of the function is gradual rather than sudden, maintenance will be able to trace the condition of the asset through the use of condition-based maintenance tasks. Figure 5.8 shows the P-F curve, which illustrates the potential and functional failure concepts together. The potential failure point (P) could be located anywhere between the initial capability and the functional failure threshold point (F). The potential failure event took place when operators measured a flow of 48 gpm. This is lower than the 50 gpm initial capability of the pump. The extension of the P-F interval, often called the *warning interval*, is driven chiefly by the nature of the function's degradation process, the methods used to measure the condition parameter, and the ability of the people performing the analysis. The P-F interval would be longer if this water flow loss were caused by impeller wear due to its mere friction with the process water, provided there were no metallurgical issues as regards chemical resistance. Also, if the operators had a super-precise flow meter with very high reading resolution, then flow degradation could be detected even earlier than by using the analog flow meter that was in place.

CLASSIFICATION OF FUNCTIONAL FAILURES

RCM-R® evaluates functional failures that are significant. Functional failures may be classified as critical, noncritical, or hidden failures. *Critical failures* are those having adverse consequences for the organization by putting the achievement of business goals at risk. Thus, functional failures affecting production capacity in any form are classified as critical, because they carry economic risks. Decreasing the speed of a production line to avoid motor overloads, for example, may be considered a critical functional failure, especially if there is no work-in-process stock to absorb the impact caused by the production rate reduction. There are failures that do not impact production capacity, but their economic consequences are still significant because of the elevated cost of repairing or replacing worn-out machinery or increases in energy consumption. When failures have the potential to put the health or safety of people at risk, they are also considered critical. If the company image or reputation is affected by the loss of a function, such an event is considered a critical functional failure. On the other hand, failure events that do not affect production capacity, people's safety, environmental integrity, or the company image at all and are not costly to repair are all considered *noncritical* functional failures. *Hidden failures* occur when asset protective devices or redundant components fail to fulfill their function. They represent the inability of an asset to carry out its hidden functions—effectively, they were already in a failed state before being called on to act. Hidden failures are (by definition) not evident to the operator or maintainer of the machine during normal operation and

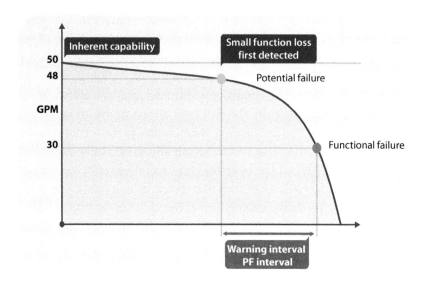

FIGURE 5.8
Scrubber pump caustic water flow P-F curve.

may expose the organization to multiple failures affecting the integrity of people, the environment, machinery, or the product. Strategies for managing failure consequences for all critical, evident, and hidden failures will be discussed in Chapter 8.

CHAPTER SUMMARY

Asset functions represent the starting point of any RCM-R® analysis once assets are chosen to undergo the process analysis. The main outcome of RCM-R® is a maintenance plan capable of sustaining the asset's functions. Thus, if functions are not properly described, the resulting maintenance program may yield less than optimal results. Assets have primary, secondary, and hidden functions, which must be described and written carefully according to the assets' operating context and performance levels in accordance with the RCM-R® process. In general, the operating context clarifies how and where the asset is going to be used. The asset owners must define the performance standard for each function of the asset. Therefore, function statements must take into consideration both the operating context and the performance level as required by the user.

RCM-R®	Functions and functional failures with classification (critical, non critical, and hidden)			
Function #	Function, performance level and operating context	#	Functional failures (loss of function)	Failures classification
2	To supply a minimum of 30 gpm of caustic water	A	Unable to supply caustic water at all	C
		B	Unable to supply the minimum required 30 gpm of caustic water	C
5	To handle maximum capacity demand at less than 75% of FLA (efficiency)	A	Unable to handle maximum capacity demand at less than 75% of FLA (amps reading > 75% of FLA)	C
11	To shut down the motor if load exceeds 125 amps (protective)	A	Unable to shut down the motor if loads exceeds 125 amps	H

FIGURE 5.9
Caustic water scrubber pump functional failures.

Functional failures occur when assets lose their ability to fulfill any of their functions. Depending on the consequences that failures may have for the organization's goals, they may be classified as critical or noncritical. Functional failures of hidden functions as the result of the loss of a protective or backup device are called *hidden failures*, because operators are not able to notice when they occur. On many occasions, unless we are actively looking for them, hidden failures do not impair the ability of the asset to perform its primary functions, and they are noticed only when a catastrophic failure takes place, and the consequences the device was intended to avoid are experienced—too late! Both functions and functional failure statements should be clearly written to be SAE JA1011 compliant. Function statements must contain a verb, an object, and a quantitative (or qualitative) performance standard whenever possible. Functional failure statements must establish the loss of the function either partially or completely. Figure 5.9 shows how functional failures for primary, secondary, and hidden functions are correctly written and classified into critical (C), noncritical (NC), or hidden (H).

6

Failure Symptoms and Causes

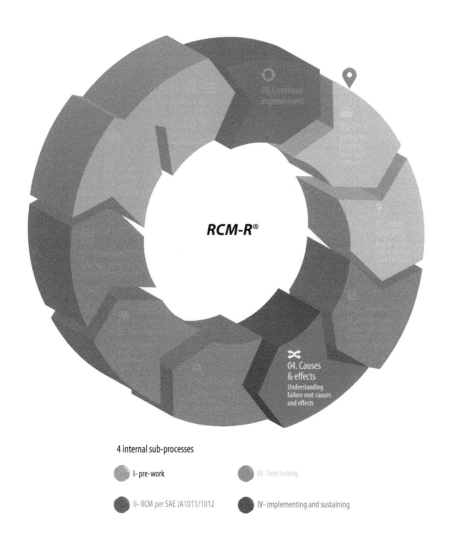

RCM-R®

04. Causes & effects
Understanding failure root causes and effects

4 internal sub-processes

I- pre-work

II- RCM per SAE JA1011/1012

III- fine tuning

IV- implementing and sustaining

RCM-R® is a consensus-driven process in which multidisciplinary teams produce strategies for managing failure consequences. Similarly, a multidisciplinary team agrees on an asset register model to help convert data into reliable information for achieving business goals. Assets are criticality-ranked by a team, who first agree on the risk matrix criteria used for this purpose. Asset functions and failures are defined by a multidisciplinary team as well. The whole RCM-R® process is consensus driven, even if the initiative to implement it first came from the maintenance department. Sometimes, however, this team-based consensus approach is forgotten, to the detriment of the analysis and the team itself.

We have seen process or operations personnel called into RCM analysis only when defining the operating context. They are the owners of the assets and are logically the ones to establish this context. But later, when defining failure modes, operators are not included because "they don't know the technical aspects (mechanical, electrical, hydraulic, HVAC, etc.) of failures and they don't understand maintenance." This is a terrible mistake. It risks transforming your RCM efforts into a futile exercise rendering far less than optimum results. Operational personnel are often the ones who first notice the failure symptoms (modes). They are close to the machines and better positioned to notice whether a noise level or pitch has changed, or whether the unusual vibration is now intermittent or continuous. Maintenance often becomes involved once failures have occurred and seldom experiences the early signs of these failures. Failure mode analysis without operations people is like getting medical attention without any direct communication with your doctor. Imagine that a friend of yours talks to the physician because he noticed you are sick. Then the doctor, based on your friend's opinions about your sickness, sends you medication, recommends therapies, requests lab analysis, and so on. No way! You (and hopefully your doctor) will not let that happen. Likewise, an effective failure consequence management strategy needs the input of operations people on the team as part of the analysis effort.

BRAINSTORMING

The step-by-step approach taken in RCM-R® helps practitioners to organize projects properly for attaining the expected goals. At this point, we have defined asset functions and functional failures. The next step is identification of failure modes and their corresponding root causes.

Brainstorming by a team is commonly used to identify failure modes and associated risks; consensus is used to finalize the list. ISO standard 31010, Edition 1.0 2009-11, recognizes brainstorming as a useful risk assessment technique. Risk assessment in general was discussed in Chapter 4.

Brainstorming entails stimulating and encouraging free-flowing conversation among knowledgeable people to identify potential failure modes and associated hazards, risks, criteria for decisions, and/or options for treatment. It involves careful facilitation to ensure that people's imagination is activated by the thoughts and statements of the others in the group and to ensure that ideas are not judged and criticized as they arise.

Effective facilitation includes stimulation of the discussion at kickoff, periodic prompting of the group, suspension of judgment, and capture of any issues arising from the discussion. Facilitation will be further explained in Chapter 12.

Brainstorming can be used as a stand-alone technique or in combination with other risk assessment methods to encourage imaginative thinking at any stage of a risk management process and at any stage of the life cycle of assets. It may also be used for high-level discussions where issues are identified, or for more detailed discussions such as failure mode identification. Brainstorming places a lot of emphasis on imagination. Therefore, it is particularly useful when identifying risks of complex and technological assets, where there is no failure data, and novel solutions will be needed to effectively mitigate risks. RCM-R® uses a consensus-driven process through brainstorming with multiskilled people who have knowledge of the system or process being assessed.

FAILURE MODES

A failure mode is an event through which a functional failure manifests itself. SAE JA1011 defines a failure mode as a single event that causes a functional failure. Failure modes can occur at the system, subsystem, or maintainable item level. They can be influenced by external factors arising both within and outside the boundaries of the asset under analysis. One of the most challenging aspects of any RCM process is the correct identification of failure modes so that they can be addressed proactively. It is often the causes of failure modes that are the targets of RCM-R® decisions. These causes are often combinations of pre-existing conditions and triggering events leading to the failure mode event. In other RCM methods, there is often confusion around exactly what a failure mode and its effects are. To help avoid this confusion and target these conditions and events most effectively, we prefer to visualize the failure mode (event) as a symptom of the functional failures (effects) rather than a cause. Why is that?

We learned from our experience and our research on RCM practices that most RCM analyses do not record failure modes clearly enough to trigger specific effective tasks to reduce the consequences of failures. All too often, failure mode statements are too vague. The descriptions often need to reflect a greater depth of understanding of the failure mechanisms. They seldom ahieve this if we attempt to identify failure modes and causes together in a single description.

For example, stating "bearing fails" is very vague. It can fail due to a variety of lubrication-related problems (too little, too much, contaminated, too hot, etc.), or due to excessive loading, or due to ingress of dirt, or due to ingress of water from washing that cleans out needed lubricants, and so on. Each of these could be considered a "cause," and each of these may even have a deeper "root cause." That one failure mode description, "bearing fails," is really an end physical effect of one or more failure mechanisms arising due to existing conditions or events in combination with conditions.

We must provide a systematic process to help RCM teams find all likely root causes of failure modes. Most people doing RCM at plants carry out their analysis after taking a 2- or 3-day introductory class. This is the way RCM is most commonly taught and applied. In their first few analyses, new practitioners are often confused by all the new jargon and nomenclature on reliability. They often make rookie mistakes such as not going deep enough into the analysis of causes of failure modes. This tendency is exacerbated if the facilitator, no matter how well trained, is also less experienced. We will discuss failure mode root causes in more detail later in the chapter.

The analysis team should identify all failure modes pertaining to functional failures that are reasonably likely to occur so as to identify the right failure management policies that will reduce their consequences to tolerable levels. Brainstorming is used to identify failure modes that are known to have happened to the asset and to other similar assets in the plant or elsewhere according to the experience of the RCM-R® team members. Work order data, as explained in Chapter 3, is also a good source of failure mode information for assets. It is also worthwhile to examine the current

PM program to identify which failure modes are already being prevented from happening. These are included in the analysis. The multidisciplinary team approach and brainstorming will also identify other potential failure modes that have not yet happened and are not being taken care of in the current PM program. If they are thought to be likely to occur in the actual operating context, then they are included in the analysis.

Failure modes can occur due to natural deterioration processes, including normal wear and tear of the asset during normal operation. Normal wear and tear (fatigue, corrosion, abrasion, erosion, etc.) will eventually cause the asset to perform below its desired performance level. The accumulation of foreign matter such as dirt, dust, or water can cause the asset to lose its ability to perform some functions. As you might expect, cement furnace blowers get very dusty. The accumulation of dust can cause the fan wheel to become unbalanced. The machine can vibrate, excessive energy is consumed to continue operating, and mechanical damage can result. Both gradual deterioration and behavior exhibited due to the accumulation of foreign matter are illustrated in Figure 6.1. Notice that the asset loses its functional capacity gradually after some time in operation.

Some human errors cause healthy machines to suddenly lose function capabilities. Chemical process pumps are often operated remotely. Unless shutoff is automated, they are sometimes inadvertently left running after emptying tanks. This can lead to failure of mechanical seals and leakage. The occurrence of this event is totally random, depending on many possible factors (which operator is on shift, quality of shift turnover, training of operators, other process upsets and distractions, etc.). It can suddenly result in an asset that was operating above its desired performance level

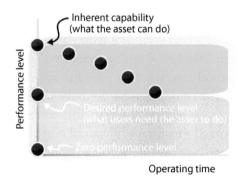

FIGURE 6.1
Gradual loss of capacity caused by deterioration.

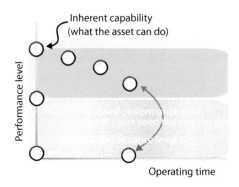

FIGURE 6.2
Sudden loss of capacity caused by a human error.

becoming completely unable to perform or losing secondary functionality, as shown in Figure 6.2.

Human errors can also result in an increase of the demand or operating performance level above the asset's inherent capability. This leads to increased applied stress on the asset. This can happen when operating speeds of manufacturing lines are increased in efforts to make up for production losses. Mistakes, such as installing the wrong mesh size filters in process lines, can result in altered flows. This can increase the operating performance level of pumping equipment beyond the asset's inherent capability.

At one client site, a process specialist engineer designed and tested a new method for attaining the required grain size of a bulk chemical powder by increasing mill speed. He proved at a laboratory scale that if the mill rotating speed increased from 4800 to 7050 RPM, then the process would yield the required grain size in a single pass instead of the three passes required using the current procedure and configuration. He changed the mill motor, sheaves, and timing belt size to meet the required speed. All looked good until the mill was tested in the mechanical shop. At the new operating speed, it ran extremely rough (vibrations) and sounded very loud. The new speed put the rotor into natural resonance—a sure-fire self-destruction mechanism. We redesigned the rotor assembly to move the rotor's natural frequency away from the new operating speed. This corrected the resonance issue, and the unit worked well for a while, until bearings wore out sooner than expected. New bearings capable of operating reliably at the new operating speed (context) had to be acquired and installed.

Increasing an asset's desired performance level beyond its inherent capacity will lead to an inherently non-maintainable situation. Maintenance (on

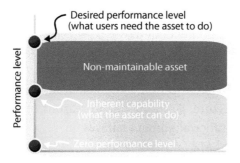

FIGURE 6.3
Non-maintainable asset.

its own) will never be able to change the inherent reliability of the asset to meet the new performance level demands. Redesign of the asset is necessary for increasing inherent capability to meet more demanding operating contexts and increased performance levels. Figure 6.3 depicts the non-maintainable situation in which the asset's inherent reliability falls below the desired performance level under the new operating context.

FAILURE MODE TYPES

Physical, electrical, chemical, and mechanical stresses, either alone or in combination, induce failures in physical assets. ISO 14224, on the collection and exchange of reliability data, calls these processes "failure mechanisms." Understanding how failure modes are initiated helps to later identify appropriate failure management policies to reduce their associated risks to tolerable levels. RCM-R® identifies the events through which failures manifest first. Then, it finds the appropriate root causes for them. RCM-R® classifies failure modes by the type of failure mechanism. It is not necessary to classify each failure mode in the analysis worksheet, but we find that consideration of the types of failure modes in the following list can be a helpful tool for generating a complete listing.

- *Mechanical:* Coded *MEC*—This includes the following subtypes:
  - General: A failure related to some mechanical defect but where no further details are known
  - Leakage: External and internal leakage, either liquids or gases

- Abnormal vibration
- Clearance/alignment failure
- Deformation: Distortion, bending, buckling, denting, yielding, shrinking, blistering, creeping, and so on
- Looseness: Disconnection, loose items
- Sticking: Seizure, jamming due to reasons other than deformation or clearance/alignment failures
- *Material:* Coded *MAT*
  - General: A failure related to a material defect, but no further details known
  - Cavitation: Relevant for equipment such as pumps and valves
  - Corrosion: All types of corrosion, both wet (electrochemical) and dry (chemical)
  - Erosion: Erosive wear
  - Wear: Abrasive and adhesive wear, for example, scoring, galling, scuffing, and fretting
  - Breakage: Fracture, breach, and crack
  - Fatigue
  - Overheating: Material damage due to overheating/burning
- *Instrumentation:* Coded *INS*
  - General failure: Related to instrumentation but no details known
  - Control failure: No regulation or faulty regulation
  - No signal/indication/alarm: No signal/indication/alarm when expected
  - Faulty signal/indication/alarm: Signal/indication/alarm is wrong in relation to actual process. Can be spurious, intermittent, oscillating, or arbitrary
  - Out of adjustment: Calibration error, parameter drift
  - Software failure: Faulty or no control/monitoring/operation due to software
  - Common cause/mode failure: Several instrument items failed simultaneously, for example, redundant fire and gas detectors; also failures related to a common cause
- *Electrical failure:* Coded *ELE*
  - General failures: related to the supply and transmission of electrical power, but where no further details are known
  - Short circuiting: Short circuit

- Open circuit: Disconnection, interruption, broken wire/cable
- No power/voltage: Missing or insufficient electrical power supply
- Faulty power/voltage: Faulty electrical power supply, for example, overvoltage
- Earth/isolation fault: Earth fault, low electrical resistance
- *External influence:* Coded *EXT*
  - General: Failure caused by some external events or substances outside the boundary but no further details are known
  - Blockage/plugged: Flow restricted/blocked due to fouling, contamination, icing, and so on
  - Contamination: Contaminated fluid/gas/surface, for example, lubrication oil contaminated
  - Miscellaneous external influences: Foreign objects, impacts, environmental influence from neighboring systems
- *Miscellaneous:* Coded *MIS*
  - General: Failure mode that does not fall into one of the categories listed above
  - No cause found: Failure mode investigated but cause not revealed or too uncertain
  - Combined causes: Several causes: If there is one predominant cause, this should be coded
  - Other: No code applicable: Use free text
  - Unknown: No information available

Let's think of a car *failing to stop* when the brake pedal is pressed. We may list many events through which the functional failure (failing to stop when the brake pedal is pressed) manifests itself. Here are some of them:

1. Brake fluid leak (Type: *MEC*)
2. Pads worn (Type: *MAT*)
3. No low brake fluid indication (Type: *INS*)
4. No DC power (Type: *ELE*)
5. Slippery or oily road (Type: *EXT*)
6. Unknown (Type: *MIS*)

All these events are failure modes, which may need further investigation to find plausible tasks for mitigating the consequences of failures to tolerable levels.

ROOT CAUSES OF FAILURE MODES

Note that the information provided by the failure mode alone may not be complete enough to find a specific failure management policy to tackle the consequences of the functional failure. Therefore, each failure mode must be further analyzed to find likely causes of its occurrence. We may find that a single failure mode has multiple root causes associated with it. Therefore, we should evaluate each cause likely to take place individually. The "5 whys" method is frequently used by the authors to assist participants in finding the most likely root cause of the failure. Most of the information and knowledge needed to carry out an appropriate analysis resides in the minds of our multidisciplinary team. It is fragmented and spread among the participants in the analysis. RCM-R® provides the means to process that valuable individual knowledge in an organized way, resulting in the identification of appropriate consequence management strategies. These strategies include proactive maintenance tasks, one-time changes, and even letting the failure occur as long as the chosen strategies are consonant with financial, safety, and environmental goals.

Root causes of failure modes can also be classified into some distinctive groups according to their nature. While it is not necessary to classify them in the RCM-R® analysis worksheet, we do find it helpful. The following list of root cause types per ISO 14224 can be used as a guide to help describe root causes appropriately:

- *Design related:* Coded *DSG*
  - General: Inadequate equipment design or configuration (shape, size, technology, configuration, operability, maintainability, etc.), but no further details known
  - Improper capacity: Inadequate dimensioning/capacity
  - Improper material: Improper material selection
- *Fabrication/installation related:* Coded *FAB*
  - General: Failure related to fabrication or installation, but no further details known
  - Fabrication error: Manufacturing or processing failure
  - Installation error: Installation or assembly failure (assembly after maintenance not included)
- *Failure related to operation/maintenance:* Coded *O&M*

- General: Failure related to operation/use or maintenance of the equipment, but no further details known
- Off-design service: Off-design or unintended service conditions, for example, compressor operation outside envelope, pressure above specification, and so on
- Operating error: Mistake, misuse, negligence, oversights, and so on during operation
- Maintenance error: Mistake, errors, negligence, oversights, and so on during maintenance
- *Failure related to management:* Coded *MGT*
  - General failure: Related to management issues, but no further details known
  - Documentation error: Failure related to procedures, specifications, drawings, reporting, and so on
  - Management error: Failure related to planning, organization, quality assurance, and so on
- *Miscellaneous:* Coded *MIS*
  - Miscellaneous: General: Causes that do not fall into one of the categories listed above.
  - No cause found: Failure investigated but no specific cause found.
  - Common cause: Common cause/mode.
  - Combined causes: Several causes are acting simultaneously. If one cause is predominant, this cause should be highlighted.
  - Other: None of the above codes applies. Specify cause as free text.
  - Unknown: No information available related to the failure cause.
- *Normal use:* Coded *AGE*
  - General: This code was added to those mentioned previously and provided by ISO 14224. Age is a very common root cause accounting for an item undergoing wear-out failures according to its inherent reliability characteristics. Examples of failure modes for which age is a plausible cause are car tire wear, air filter clog, pump impeller wear, valve seat wear, and so on.

HOW MUCH DETAIL?

How does the team know it has found a real root cause? How many times do they need to ask "why?" Well, root causes of failure modes are described in the form of a statement. Bearing wear is a very typical failure mode found in almost every rotating component. Technically speaking, maintenance tasks will be applied to root causes and not to symptoms (failure modes). Consider, for instance, a centrifugal pump bearing wear situation. If we were to determine possible causes for it, and we used the "5 whys" technique, we might begin with the following logical question and get the following sequence of possible and plausible answers:

- Why did the bearing wear (failure mode)?
  - It was due to a lubrication problem. Why?
  - It had insufficient grease. Why?
  - The grease line diameter was too small. Why?
  - Due to an error in design. Why?
  - Due to the use of the wrong specification. Why?
  - Due to a human error. Why?
  - Due to a distraction. Why?
  - Because the designer is in love. Why?
  - Because he found the right partner and will get married. Why?
  - Because he is Catholic. Why?
  - Because his family is …

Clearly, we can take this very far. In fact, every event, such as a failure mode, is but one in an infinite series of events, some of which we can probably control and some we cannot. If we go into too much detail or too far back along the chain of events, we have clearly gone too far; we need less detail. We also want to avoid overworking the problem. If we just describe the failure mode alone (i.e., bearing wear), then the statement is too wide to enable us to find a single appropriate task; we need more detail.

We are certain that we arrived at a reasonable root cause if we are able to find a consequence management policy capable of reducing the failure risk to a level we can tolerate. It is also clear that our RCM-R® analysis team can't be too detailed or superficial when establishing the failure mode–root cause statements. We only need to deal with one event in the infinite series of events to change the rest of the series, but which event do we choose? In the example of bearing wear, different team members may have different ideas on which of the root causes is the one to deal with.

A pump operator on the team may consider the answer about incorrect design to be sufficient. If a designer is on the team, he may prefer to identify the use of an incorrect specification that was called for. A maintainer might consider insufficient grease to be the root cause. Who's right?

In fact, they are all right, but which of the possible solutions will be the most workable for the organization in its present operating context? Do we change the design, change how projects are specified, or change the amount of grease applied? The RCM-R® facilitator must ensure that the analysis team stays focused on providing just the necessary information to get to a workable solution. The facilitator's role includes appropriate time management around issues such as this. It is important to do the right amount of work, not too much or too little, in arriving at good consequence management strategies.

As thorough as our above example appears, it is also incomplete! We all know that wrong design is not the only root cause for a centrifugal pump bearing wear. The bearing may have been installed incorrectly, leading to premature wear. This failure mode root cause may be brought to the analysis team by the pump owner or the maintenance representative. Wrong installation is a fairly common cause, and the owner of the asset may be interested in understanding the reason(s) for it. The "5 whys" analysis may begin with the same question and lead the team to a different (additional) plausible root cause as follows:

- Why did the bearing wear (failure mode)?
  - The bearing was installed incorrectly by the contractor. Why?
  - The installer made an error. Why?
  - He used the wrong installation practice. Why?
  - He lacked the appropriate tools to do the job correctly. Why?
  - …

If the analysis team identified wrong contract pump installation as a possible cause of bearing wear, they may review the contractor's qualifications for the job to prevent this situation from happening in the future. Perhaps, the contractor even used a carpenter or a pipe fitter who isn't trained in fitting practices to align the pump shaft without the use of precision tools. This sort of insight into what happens is actually a fairly typical insight gained by analysis teams from their own members. The information comes to light because we are asking knowledgeable and experienced team members questions that are relevant to events they've seen occur.

CHAPTER SUMMARY

Failure modes are events through which failures show up. RCM-R® considers failure modes to be symptoms of failures for which root causes must be identified. Functional failures may manifest through many failure modes, which, in turn, may be produced by multiple root causes. A pump may fail to provide the needed flow (failure) due to a broken coupling (failure mode) provoked by a wrong installation caused by lack of training (root cause). RCM-R® classifies failure modes by their failure mechanisms per ISO 14224, recognizing mechanical, material, instrumentation, electrical, external, and miscellaneous types of failure modes. Each failure mode is further analyzed to find reasonable root causes that can be classified further as design, installation/fabrication, operation/maintenance, management, and miscellaneous failure mode root cause types as identified in ISO 14224. While these classifications are not essential to the process, they are helpful in making sure the analysis team hasn't missed something important. The complete failure mode and root cause analysis must be carried out by a multidisciplinary team with experience on the process and the asset under analysis. RCM-R® facilitators apply formal brainstorming techniques for stimulating team discussion and ideas, and help the team identify failure modes and their root causes to the right level of detail to enable later determination of appropriate failure consequence management strategies.

Figure 6.4 shows how failure modes and root causes for functional failure A of function #2 of the caustic water scrubber pump example are correctly written and classified for analysis.

There may be plenty of failure modes associated with this functional failure, and the analysis team must identify all the likely failure mode events and their root causes. So far, we have answered the first three basic questions of RCM:

1. *What are the functions and associated desired standards of performance of the asset in its present operating context* (functions)?
2. *In what ways can it fail to fulfil its functions* (functional failures)?
3. *What causes each functional failure* (failure modes root causes)?
    In the next chapter, we determine each failure's impact by describing the sequence of events that happens when each failure mode occurs, answering the fourth question in the process:
4. *What happens when each failure occurs* (failure effects)?

RCM-R®

Functions and functional failures with classification
(critical, noncritical, and hidden)

Function #	Function, performance level and operating context	#	Functional failure (loss of function)	Failures classification		Failure modes/ types & root causes/ type					
					#	Failure mode	Type	#	Root causes	Type
2	To supply a minimum of 30 gpm of caustic water	A	Unable to supply caustic water at all	C	1	Pump coupling wear	MAT	c	Misalignment caused by wrong practices	MGT
					2	Pump cavitation	MEC	a	Low suction pressure due to a dirty strainer	AGE

FIGURE 6.4
Caustic water scrubber pump failure modes and root causes example.

7

Quantifying Failure Impacts

RCM-R®

10. Continual improvement

04. Causes & effects
Understanding failure root causes and effects

05. Detection
Understanding how failures are detected

4 internal sub-processes

- I- pre-work
- III- fine tuning
- II- RCM per SAE JA1011/1012
- IV- implementing and sustaining

We have explained how to describe asset functions, functional failures, failure modes, and root causes in the preceding chapters. At this stage of an RCM-R® project, the multidisciplinary analysis team, by way of its participants' knowledge, has documented in detail all of the asset's primary, secondary, and hidden functions along with their corresponding functional failures. Failure modes and root causes may have also been identified and classified by the team. By this point, we find that the overall asset knowledge of the team members has also improved dramatically by virtue of the information shared so far. The whole process makes the machine operator more knowledgeable about failures, their causes, and maintenance. Operational- and maintenance-induced failure modes may have been better understood too. Basically, the team understands what the asset must accomplish, how it fails to do so, and the reasons for these failures.

Now, we need to understand failure relevance with respect to impact on business goals. RCM-R® assesses failure causes and prioritizes their corresponding failure management policies according to their criticality. The analysis team clearly describes what will happen when each failure mode with its corresponding root cause actually occurs. Each failure effect must link to an individual root cause. Several causes may often have the same effect, but they should be evaluated separately, because later, the decisions on what to do to manage them are made failure cause by failure cause.

GUIDE QUESTIONS FOR DESCRIBING FAILURE EFFECTS

The description of failure effects enables us to justify the type of consequence management policy for risk elimination or reduction to a level that the system owner can tolerate. It should also include all the information needed to support the evaluation of the failure. To do this, the RCM-R® team assumes that no proactive maintenance is done to prevent the failures from happening when it sets out to describe the failure effects. The team must avoid considering existing maintenance tasks at this time. When assessing failure effects, the team considers the operating context and performance level of the asset. Compensating provisions that may be built into the asset design or its operating procedure, which are intended to eliminate or reduce the consequences of the failure, should be accounted for when describing the effects and assessing the consequence severity of failure mode root causes. This includes standby devices, work-in-process stock, spare parts availability, safety devices, procedural actions, and so on. The failure effects should describe and quantify the impact of every root cause on business objectives with regard to cost, safety, and environmental impacts. Other impacts may also be important. For instance, deterioration of the firm's reputation resulting in potential business loss may be a risk in some cases.

The following questions may be used as a guide to construct a statement of effects (in the form of a descriptive paragraph):

1. What are the facts proving the failure occurrence (evidence)? Where the failure is hidden, what would happen if a multiple failure took place?
2. How is the safety of the people around the failed asset affected?
3. How are environmental goals impacted?
4. How is production or the operations affected by the failure?
5. What kind of physical damage is caused by the failure? How costly is the failure in terms of maintenance and repair?
6. Is there any secondary damage? What must be done to restore operations? How long would it take?
7. How likely is the failure to occur? Has it happened before?

RCM-R® facilitators guide the team to formulate precise statements of effects for every failure root cause. This is not always an easy task to complete, and it may require the gathering of more information from the field if the team on its own cannot completely describe what happens.

HOW IS THE FAILURE DETECTED?

Answering Question 1 requires mentioning whether or not there has been a change in the behavior of the asset prior to the failure. Did temperature, speed, vibration, or noise level change before the failure took place? Is the operator somehow warned by an alarm sounding when the failure occurs? This question is closely related to an important feature of RCM-R®. The facilitator should explicitly request the analysis team to explain how the failures are discovered or detected. Some failure mode causes are relatively easy to spot, and others are not, even requiring diagnostic work to isolate them. RCM-R® offers some straightforward guidance to simplifying the approach a bit. Participants should identify how the failure mode root cause is detected by selecting one of the following answer choices:

- *Through senses*: Mostly visually but also by ear, smell, or touch.
- *Through alarm*: Including audible alarms, warning lights, and so on.
- *Operator*: The asset operator has the knowledge and skills to identify the failure mode root cause.
- *Internal expert*: In-house craftspersons, such as mechanics, electricians, or specialists in reliability, quality, or environmental health and safety (EHS), have the required knowledge and skills to identify the failure mode root cause.
- *External experts*: In-house personnel lack the expertise and skills to identify the failure mode root cause. Thus, specialized contractors have to be brought in to discover the failure root cause.

Examples of statements answering Question 1:

"When the pump strainer clogs the motor stops and the horn alarm sounds."

"If the acetone tank safety relief valve is defective and there is an overpressure, then the tank will collapse."

"When the impeller is worn enough, the low flow alarm will blink, but the operator is not able to know the cause of the failure yet. Mechanics confirm the root cause of the failure on asset disassembly."

HOW IS THE SAFETY OF THE PEOPLE AROUND THE FAILED ASSET AFFECTED?

Question 2 must be answered precisely, because enterprise safety goals may be put at risk if this important aspect is not properly assessed. RCM-R® carefully documents, evaluates, and ranks safety risks by answering this question. We provide guidance in the form of these answer choices to facilitate the description and eventual assessment of the safety risks posed by each failure cause:

- *No impact*: Accident likelihood is negligible.
- *Minor safety impact*: Injuries not requiring medical treatment may occur.

- *Moderate safety impact*: Injuries requiring medical treatment may occur.
- *Severe safety impact*: Serious personnel injuries may occur.
- *Catastrophic safety impact*: Loss of lives may occur.

Specific safety threats posed by the failure causes must be clearly described. The following examples may be helpful for describing some common situations:

"Steam leaks may cause second degree skin burns."
"Caustic soda escape may cause chemical burns requiring major surgery."
"Exposure to live cables may result in electrocution."
"Sharp edges may cause serious injuries requiring stitches."
"May cause pacemaker malfunction and death."

HOW ARE ENVIRONMENTAL GOALS IMPACTED?

Responsible organizations put people and the environment first! Also, they strive to recognize, evaluate, and mitigate potential environmental hazards. Some of these environmental threats pose other collateral damage, including high fines for not complying with applicable environmental standards, possible halting of production, and reputation deterioration that could trigger future business loss. Quantifying environmental impact is sometimes unfair to the environment itself, as the impact can be both physical and financial. We recall evaluating an RCM project in which all environmental impacts were described in terms of the dollar amount of the fines that would be levied. Question 3, on failure effects, may be answered in terms of the severity of the event's environmental impact followed by a brief description of the incident itself. The following guidance facilitates the description and eventual assessment of the environmental risks posed by each failure cause:

- *No environmental impact*: Environmental threat likelihood is negligible.
- *Minor environmental impact*: Minor pollution is possible.
- *Moderate environmental impact*: Some pollution exceeding applicable standards and practices is possible.

- *Severe environmental impact*: Significant pollution according to applicable standards and practices may occur.
- *Catastrophic environmental impact*: Extreme pollution and environmental damage may occur.

Specific threats posed by the failure causes to the environment must be clearly described. The following examples may be helpful for describing some common situations:

"Oil seal leaks are contained in the pump pit, causing negligible environmental threat."

"Collapsed underwater oil line may cause severe contamination and threat to marine life."

"Broken air purifier filter may cause air contamination, seriously affecting the health of endangered bird species."

HOW IS PRODUCTION OR OPERATIONS AFFECTED BY THE FAILURE?

Some failures have direct impacts on production. Their failure effects statements should indicate how production is affected and for how long. It is wise to measure the failure impact in terms of units of production deferred as the result of the failure. Production time loss may be converted into actual units of production or even into the cost of the postponed production. It is recommended that teams evaluate the most likely worst-case scenario.

Let's say that we have a shared backup unit. We have Pump A and Pump B on duty and Pump C backing up both A and B units. It is recommended that we consider the case of an on-duty pump failure when the shared backup pump is unavailable. In other words, we need to consider that both on-duty pumps may be at a failed state simultaneously at some point. The failure effects on operations may be measured in terms of *downtime* (hours), *raw materials* lost, *production volume* deferred, *loss of throughput* due to machine speed reduction, or possible loss of *future contracts* due to lack of product quality or for noncompliance with contractual production supply. All of the above may be converted into the language every business manager understands: money loss. We facilitate the failure statement development by considering the following levels of production impacts when answering failure effects Question 4:

- *No loss of production*: Production is not affected in any form.
- *Minor production loss*: Production loss may be recoverable and does not impact business sales goals.
- *Moderate production impact*: Production loss just below acceptable limit. Uncorrected situation may lead to unrecoverable losses.
- *Production loss above acceptable limits*: Downtime, speed reduction causing significant throughput drop, substantial quality defects affecting reputation or gain of potential business, and so on.
- *Extensive stop in production*: Excessive downtime, throughput drop, quality defects and loss of potential business, and so on.

Failure effects statements must clearly describe in which ways production is affected both at present and in the future. The following examples may be helpful for describing some common situations:

"Filler speed must be reduced to 70% to avoid production stop resulting in X amount of production deferment."
"The defective motor is replaced during the maintenance shift."
"The mechanical seal replacement takes 3 hours, leading to an actual delay of 5 hours of production representing X volume of production loss."
"All raw material in the tank is lost when the heating system is not able to regulate the temperature, causing a 4 hour production delay."

WHAT KIND OF PHYSICAL DAMAGE IS CAUSED BY THE FAILURE? HOW COSTLY IS THE FAILURE IN TERMS OF MAINTENANCE AND REPAIR?

These questions pertain to possible physical damage that shows up when the failure occurs. Is there material deformation or distortion? Did the tank roof collapse? Did the shaft break? This is the type of information sought by this question.

Almost all failures represent an actual monetary cost due to repair activities that must be estimated and documented. Those costs are relevant to our decision-making process on whether or not to actually take proactive steps. An on-duty pump may have stopped because of bearing wear due to age. Then, the backup unit may have taken over service without causing any production loss. But, somebody's budget is still affected, most likely maintenance. Cost estimation should include material, labor, spares, contractor, and logistic costs, among others, associated with the repair activity.

IS THERE ANY SECONDARY DAMAGE? WHAT MUST BE DONE TO RESTORE OPERATIONS? HOW LONG WOULD IT TAKE?

Asset failures may cause malfunctions or increased operating costs in other systems that are outside of the analysis boundaries. For example, excessive air leaks in an instrument air component on a particular asset may cause increased power consumption by air compressors. This leads to higher energy costs. If there is an experienced ultrasound inspector within the analysis team, he/she will be able to estimate cost based on typical leak rate scenarios. In one case, an air leak survey was carried out at an organic synthesis chemicals plant for the first time in 20 years. The results were astonishing. They were wasting more than 45% of the air they produced just because of air leaks throughout the distribution line. The plant was running well, and production volumes were met, but at the added expense of excessive air compression cost. After learning this, the company corrected the leaks and was then able to shut down one of the four in-line compressors, saving a great deal of energy and dollars.

Sometimes, more than just repair activity is needed to restore operations after a failure occurs. Process line passivation, quality control checks, and operator pre-start checklist, among others, must take place to enable operations to resume production. All of these activities should be outlined in the failure effects statements. Here, we can see the importance of the multidisciplinary team approach, through which multiple aspects of what has to be done are visualized and described. It is not just about maintenance. The time and cost related to the operation, quality, and repair activities needed to put assets back into service should be considered and recorded in the failure effects statements.

Care must be taken to avoid confusing downtime with repair time. Account for all the time the asset was not able to produce at the expected rate, not just the time it was being repaired. Thus, all of the time from stoppage till production resumes and reaches "cruise speed" should be recorded as part of the failure effects statement.

DOCUMENTING FAILURE EFFECTS STATEMENTS IN THE RCM-R® WORKSHEET

Our multidisciplinary approach provides a comprehensive view of the effects that each failure carries for the organization. We are constructing a comprehensive failure mode effects and criticality analysis (FMECA) catalog for the asset and for future use. Outside the RCM analysis itself, this catalog can be consulted and used as a source of information for root cause analysis when failures occur. We will see later how the catalog is expanded beyond the FMECA analysis to include classification of failure effects according to consequences for which failure management policies are recommended, but for now, let's look at how we document the failure effects statements.

All seven failure effects questions must be addressed. Documentation must be very specific. When a failure does not have an impact on production or the environment, for example, we say so. When it does, we describe how and how much. RCM-R® is a formal tool used by professionals not only to optimize operational and maintenance costs but also to support the organization in attaining its overall business goals, including sales, environmental, safety, energy management, and so on. The authors encourage teams to provide thorough and complete documentation throughout the RCM-R® analysis process to enable the greatest value to be derived from it.

Let's go back to our scrubber pump of Figure 5.2 and expand its failure mode analysis to include the failure effects statements for two failure root causes. Figure 7.1 considers all the information requested by the seven guide questions for describing the failure effects statements for two of the pump's failure causes.

ISO STANDARD–BASED FAILURE EFFECTS RISK ANALYSIS

The reader may have realized that getting to this point of an RCM-R® analysis takes a lot of work. You may be wondering whether it is necessary to

RCM-R®

Functions and Functional Failures with Classification
(Critical, Non-Critical, and Hidden)

Function #	Function, performance level, and operating context	#	Functional failure (loss of function)	Failures Classification	#	Failure mode	Type	#	Root causes	Type	Failure effects*
2	To supply a minimum of 30 gpm of caustic water	A	Unable to supply caustic water at all	C	1	Pump coupling wear	MAT	c	Misalignment caused by wrong practices	MGT	Vibration alarm sounds. Mechanic investigates, finds coupling wear. No safety risk. Environmental alert posted. Operations shut down for 1 hour for repair. Costs: repair $100, downtime $500k. Chance that a longer repair results in loss of material in reactors (+$300k). Misalignment is common problem increasing ongoing energy costs 10% and reducing coupling life to 8000 hours.
					2	Pump cavitation	MEC	a	Low suction pressure due to a dirty strainer	AGE	Low pressure indication in HMI. No safety or environmental risk. Strainer replacement takes 10 mins and costs $50. Production not disrupted. Historic time for strainer to plug is 300 hours.

Failure Modes/Types & Root Causes/Type

*What happens when the failure occurs?
1. What are the facts proving the failure occurrence?
What would happen if a multiple failure, in the particular case of a hidden failure, takes place?
2. How is the safety of the people around the asset affected?
3. How are environmental goals impacted?
4. How is production or the operation affected by the failure?
5. What kind of physical damage is caused by the failure?
How costly is the failure in terms of maintenance and repair?
6. Are there secondary damages? What can be done to restore operations? How long would it take?
7. How likely is the failure to occur? Has it happened before?

FIGURE 7.1
Partial gas scrubber water circulation pump failure modes and effects analysis.

evaluate all of the failure causes to assign a maintenance or redesign task to the failure. The authors are glad to answer: NO!

RCM-R® is a process to identify and manage risks to assign plausible risk management policies to eliminate or reduce their consequences to levels tolerable to the owner of the asset. Therefore, a risk analysis for all of the likely failure causes is conducted to prioritize their recommended actions. This failure effects screening exercise often results in filtering out 30%–50% of identified failure causes. This means that no proactive or one-time action tasks are recommended for them. The team's time is spent on task assignment analysis for the failure causes posing significant risk to the operations. But, what is considered a significant risk?

Organizations of all types and sizes face internal and external challenges that make it uncertain whether they will achieve their objectives. The effect this uncertainty has on an organization's objectives is "risk" as per ISO Standard 31000 (2009), "Risk Management—Principles and guidelines."[1] All the activities of an organization involve risk. Organizations manage risks by identifying, analyzing, and mitigating them. They evaluate whether the risk should be modified by risk treatment to satisfy their risk criteria. Throughout this process, they communicate and consult with stakeholders, and they monitor and review the risk and any controls that are modifying the risk. They are determining whether or not further risk treatment is required. RCM-R® is applied during the design and operational phases of the asset's life cycle, during which vital activities involving significant risks are performed. The sooner risks are identified and evaluated, the more efficient our asset management and the greater the value realized from our assets. Organizations determine how best to manage risk by identifying and analyzing it. This is what the standard is all about. The concepts in the standard can be applied to any type of risk, whatever its nature, whether it has positive or negative consequences, bearing in mind that not all risks are bad. However, when it comes to equipment or system failures, the risks are generally viewed negatively.

The risk management process should be an integral part of management, embedded in the organization's culture and practices, and tailored to the business processes of the organization. RCM-R® is an excellent tool for identifying and dealing with risks arising from equipment failures. The general risk management process is shown in Figure 7.2.

RCM-R® uses a process to evaluate the risks of failure effects that is very similar to that shown in Figure 7.2. Each of the five stages of the risk management process (marked with numbers 1 through 5 in Figure 7.2) has

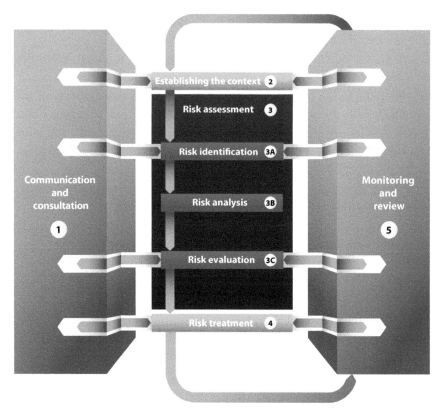

FIGURE 7.2
Risk management process per ISO 31010 (2009).[2]

a counterpart in the RCM-R® process. Stage 3 is the actual risk assessment process in which risks are identified, analyzed, and evaluated.

Risk identification (3A) entails the identification of the sources of risk, areas of impacts, events (including changes in circumstances), their causes, and their potential consequences. The aim is to generate a comprehensive list of risks based on those events that might create, enhance, prevent, degrade, accelerate, or delay the achievement of objectives. The organization should apply risk identification tools and techniques that are suited to its objectives and capabilities as well as the risks faced. Relevant and up-to-date information is important in identifying risks.

Risk analysis (3B) involves understanding how the risk develops. Risk analysis provides an input to risk evaluation, decisions on whether risks need to be treated, and the most appropriate risk treatment strategies and methods. Risk is analyzed by determining consequences, their likelihood, and other attributes of the risk. Analysis can be qualitative, semiquantitative, or quantitative, or a combination of these, depending on the circumstances.

Risk evaluation (3C) entails making decisions, based on the outcomes of risk analysis, about which risks need treatment and the priority for treatment implementation. Risk evaluation involves comparing the level of risk found during the analysis process with risk criteria established when the context was considered. Based on this comparison, the need for treatment can be considered.

RCM-R® FAILURE EFFECTS RISK ASSESSMENT MATRIX

RCM-R® evaluates the relative criticality of each individual root cause to decide risk treatment methods accordingly. All failure causes' effects are evaluated by using a risk matrix comprising three components: *severity*, *likelihood*, and *detectability* of the failure. Our three-dimensional matrix weighs the combined potential impact of these three aspects on business objectives. A multidisciplinary team must agree on risk matrix criteria, as happens with asset criticality ranking, as explained in Chapter 4. Then, a risk number (also called the *risk priority number* [RPN]) is obtained for every failure effect when the team decides on the magnitude of each risk matrix component. As a result, failure effects are ranked (Figure 7.3).

Figure 7.4 shows a failure effects risk assessment matrix. The team follows the same approach as that used for ranking assets by criticality, but now, the information needed for the analysis is already contained in each failure effects statement. Because of this readily available information, the exercise becomes fairly easy and straightforward.

CHAPTER SUMMARY

Failure effects statements establish what happens when failures occur. RCM-R® provides some key questions the analysis team must answer to ensure that failure statements are formulated correctly. The questionnaire asks for specific details on failure detection techniques, safety risks, environmental issues, production impact, maintenance costs, and the likelihood that the events causing the failure will actually occur.

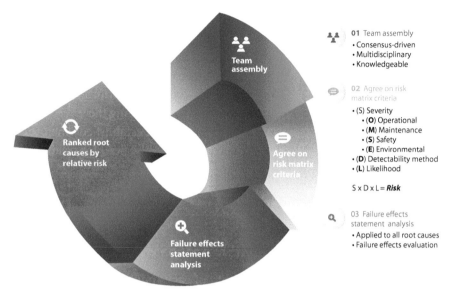

01 Team assembly
• Consensus-driven
• Multidisciplinary
• Knowledgeable

02 Agree on risk matrix criteria
• (**S**) Severity
    • (**O**) Operational
    • (**M**) Maintenance
    • (**S**) Safety
    • (**E**) Environmental
• (**D**) Detectability method
• (**L**) Likelihood

S x D x L = **Risk**

03 Failure effects statement analysis
• Applied to all root causes
• Failure effects evaluation

FIGURE 7.3
RCM-R® failure effects risk assessment model.

The effects that the impact of failures will have on the organization's goals are visualized and evaluated following guidelines outlined in International Standard ISO 31010:2009 for risk management techniques. Each failure effects statement is then used to determine the risk ranking using the failure effects risk evaluation matrix (Figure 7.4). Risk numbers enable the analysis team to rank them according to their relative criticality. This is used as a screening tool for deciding which failure causes require no mitigation at all and for selecting and prioritizing failure consequence management policies (Chapter 10). Figure 7.5 shows how risk numbers are documented in the RCM-R® worksheet. Note that the three criteria (severity, likelihood, and detectability) are chosen using the table shown in Figure 7.4 considering the failure effects statements for each failure root cause.

The example shown in Figure 7.5 presents a significant difference between the values obtained for two failure causes, both for the same functional failure. The coupling failure issue due to misalignment impacts production with some $500,000.00 in delayed production. It can only be detected by the maintenance technician with a yearly frequency, and its risk assessment produces a value of over 270. The second failure cause, cavitation due to the dirty strainer, has no significant production delay

Class	Safety	Environmental	Production	Maintenance	No known events	Failure known to occur	Failure has occurred in plant	About 1 failure per year in systems	Several failures per year in systems	Visual	Alarmed	Operator detects the failure mode	Plant specialist detects the failure mode	External specialist is needed to detect the failure mode
A	X2 no effect	X1.5 no pollution	X1.0 no stop	X0.5 no cost	0	0.4	0.8	1.2	1.6	1	2	3	4	5
B	X4 *Injuries not requiring medical treatment * No effect on safety function	X3.5 minor pollution	X3.0 X < 2% of plant capacity	X2.5 maintenance Cost X < $25,000.00	2	2.4	2.8	3.2	3.6	2	3	4	5	6
C	X6 *Injuries requiring medical treatment *Limited effect on safety function	X5.5 some pollution	X5.0 2% < X < 20% plant capacity	X4.5 maintenance cost 25K < X < $50K	4	4.4	4.8	5.2	5.6	3	4	5	6	7
D	X8 *Serious personnel injury * Potential for loss of safety functions	X7.5 significant pollution	X7.0 20% < X < 50% plant capacity	X6.5 maintenance cost $50K < X < $500K	6	6.4	6.8	7.2	7.6	4	5	6	7	8
E	X10 *Lost of lives *Vital safety-critical systems inoperative	X9.5 major pollution	X9.0 X > 50% of plant capacity	X8.5 maintenance cost X > $500K	8	8.4	8.8	9.2	9.6	5	6	7	8	9

FIGURE 7.4

RCM-R® failure effects risk evaluation matrix.

Functions and Functional Failures with Classification
(Critical, Non-Critical, and Hidden)

RCM-R®

Failure Modes/Types & Root Causes/Type

Function #	Function, performance level, and operating context	#	Functional failure (loss of function)	Failures classification	#	Failure mode	Type	#	Root causes	Type	Failure effects*	Severity	Likelihood	Detectability	Total
2	To supply a minimum of 30 gpm of caustic water	A	Unable to supply caustic water at all	C	1	Pump coupling wear	MAT	c	Misalignment caused by wrong practices	MGT	Vibration alarm sounds. Mechanic investigates, finds coupling wear. No safety risk. Environmental alert posted. Operations shut down for 1 hour for repair. Costs: repair $100, downtime $500k. Chance that a longer repair results in loss of material in reactors (+$300k). Misalignment is common problem increasing ongoing energy costs 10% and reducing coupling life to 8000 hours.	9	9.6	8	691.2
					2	Pump cavitation	MEC	a	Low suction pressure due to a dirty strainer	AGE	Low pressure indication in HMI. No safety or environmental risk. Strainer replacement takes 10 mins and costs $50. Production not disrupted. Historic time for strainer to plug is 300 hours.	0.5	1.6	2	1.6

*What happens when the failure occurs?
1. What are the facts proving the failure occurrence?
What would happen if a multiple failure, in the particular case of a hidden failure, takes place?
2. How is the safety of the people around the asset affected?
3. How are environmental goals impacted?
4. How is production or the operation affected by the failure?
5. What kind of physical damage is caused by the failure?
How costly is the failure in terms of maintenance and repair?
6. Are there secondary damages? What can be done to restore operations? How long would it take?
7. How likely is the failure to occur? Has it happened before?

FIGURE 7.5
RCM-R® Caustic soda pump failure effects risk evaluation matrix example.

and costs only $50.00. Its score was only 2.4 points. Later, the team will put more effort into failure consequence management policies for the misaligned coupling than it will for a plugged strainer.

REFERENCES

1. International Standard IEC/ISO 31000:2009, *Risk management: Principles and guidelines*, ISO copyright office, Case postale 56, CH-1211 Geneva 20.
2. International Standard IEC/ISO 31010:2009, *Risk management: Risk assessment techniques*, ISO copyright office, Case postale 56, CH-1211 Geneva 20.

8

Overview of Maintenance Strategies

OVERVIEW OF STRATEGIES FOR MANAGING FAILURE CONSEQUENCES

Managing failures and their consequences requires action. RCM-R® is a decision-making tool. It produces actionable decisions, each being specific to the failure modes analyzed. Those actions include both recurring tasks and one-time changes.

Recurring tasks are defined in sufficient detail to ensure they are executed as intended by the analysis team. They include the specification of who should do the task (maintainer or operator) as well as the task frequency.

Maintainer tasks generally require the use of skills, knowledge, or tools that are not normally available to the operators. Maintenance tasks include both disruptive interventions as well as nonintrusive monitoring activities.

Operator tasks are usually in the form of monitoring, minor adjustments, performance testing, and basic machine care activities. Technically, those tasks are forms of maintenance work; however, they do not require deep levels of trade expertise, and they are generally convenient for operators to carry out during normal operational activities. These "basic care" tasks generally don't require the use of tools. They include lubrication, cleaning, and using the human senses to determine whether things are running "normally."

To illustrate the difference between tasks designated for an operator and for a maintainer, consider the case of your family car, where we have both types of tasks. As the owner and operator, you make sure that working fluids (e.g., oil, transmission fluid, and brake fluid) are topped up. You make sure that consumables are replenished (e.g., windshield fluid and fuel). You keep the vehicle clean, and in washing it, you notice minor scratches, dents, or other blemishes on surface finishes. As you operate it, you notice that the windshield wipers are doing their job when needed, that your brakes are working, that power delivery is steady and strong, that heating or air conditioning is working, that engine noise and passenger compartment noise are "normal," that the engine isn't overheating, and that you have enough fuel. You probably check your tires for wear and make sure they are inflated to the correct pressure. If you are diligent, you will also make sure your lights, brake, and turn signal lights all function correctly and that your spare tire is inflated properly. These are all tasks that fall into the category of "basic care"—care by the owner/operator. More complex work normally involves professionals. Your auto-mechanic will carry out most repairs that are needed, change your oil and transmission fluids, rotate and change your tires, check your car's electronic systems on a diagnostic machine, inspect and change your brake pads and rotors, and so on. Those tasks require more skill, knowledge, and tools that most owners don't have or know how to use.

In RCM-R®, we consider the skills, knowledge, and ability of operators and maintainers, and we make decisions about who should do the work considering those factors.

Running the asset to failure and making one-time changes to training, procedures, or design are also valid and typical outcomes of RCM-R® decision-making. More on these later.

TECHNICALLY FEASIBLE AND WORTH DOING

Each decision deals with the causes of failures—case by case. The solutions we choose must be technically capable of dealing with the failure causes, and they must be worth doing from a perspective of costs or risk mitigation. This means that the task must be able to deal with the characteristics of the failure itself and that the task must reduce the risks or costs of the failure consequences to levels that are tolerable.

Recurring maintenance strategies are tasks that are carried out on a repeated basis, usually at a fixed frequency or task interval. They are carried out by maintainers or operators as appropriate. Recurring strategies include predictive, preventive, and detective maintenance (DM).

PREVENTIVE MAINTENANCE

Preventive maintenance is used where we have a known end of the useful life of the asset. By taking action before we reach the end of useful life, we *prevent* its occurrence; hence the name. Preventive action is carried out as close as possible to the end of useful life but before the failure occurs. Its timing is driven by time or usage of the asset since its last intervention. Time-based maintenance is the most common, and as such, preventive maintenance is often referred to as time-based maintenance (TbM). Here, we will use that acronym.

In Chapter 2, we presented a brief history of RCM. We described the six patterns of conditional probability of failure that Nowlan and Heap identified in their landmark study. Where failures occur as an item ages or as it is used (failure patterns A, B, and C), we can forecast the end of the item's useful life and plan to restore or replace it before the failure occurs. Using statistical failure data analysis (i.e., Weibull analysis), we can accurately determine the probability of failure up to any point in time (i.e., hazard). We can time replacement or restoration activity before we reach a point where the probability is beyond our level of tolerance. Note that even with preventive actions, there will still be a probability of failure before we act. For example, we may choose to replace a component that wears, corrodes, or erodes when it has only a 10% cumulative probability of failure to prevent 9 out of every 10 possible failures. We could also choose to act at two standard deviations (2σ as shown in Figure 8.1) before mean time between failures (MTBF) or some other age we are comfortable with.

TbM comprises replacement or restoration activities, both intended to return the asset/item to the "as new" functional condition, or as close to it as practical. Replacement with a new item certainly achieves this. Restoration work isn't quite as dependable, but it is usually less expensive, making it an attractive option. There are exceptions, too. In some cases, restoration work may be done with new surface coatings—enhancing surface resilience or using new materials that are more resistant to failure. These improvements on the original design are "design changes," and they work in conjunction with the preventive action. For example, replacing a bronze-nickel pump impeller with one that is ceramic coated may lead to more resistance to erosion in slurry service, thus extending its useful life. In cases like this, the design change (often implemented entirely by maintenance) may lengthen the time between preventive interventions,

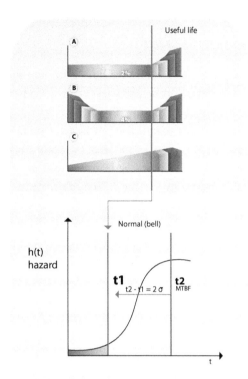

FIGURE 8.1
Useful life for age or usage failure modes.

but it probably won't eliminate them entirely. The design change results in an extension to the useful life and to the age at which the redesigned component will ultimately need to be replaced.

The major drawback with TbM is that it usually requires the machine to be shut down and disassembled to gain access to the failing item. Where continuous operation or very high availability is needed, this imposes the need for unwanted downtime. Another downside to TbM is that disassembly intrudes on other components or parts that are not normally subject to failure pattern A, B, or C. In those cases, this intrusive intervention can inadvertently cause other failures or increase the probability of their occurrence, especially if they exhibit pattern F (infant mortality) characteristics.

For example, the replacement of reciprocating compressor valves entails the removal of piping, cylinder heads, and replacement of gaskets, all of which often fail as infant mortality (pattern F) failures. Replacing them at a fixed interval increases the overall probability of failure. Opening up the cylinder head may expose lubrication or cooling fluid channels, thus creating a period of increased risk of contamination of those working fluids and triggering random failures elsewhere in the machine later. No matter how careful we are, mechanics are human, and like all of us, they may make mistakes. Unfortunately, the reciprocating compressor valves have a very distinct usage-related failure characteristic (pattern A or C), and their replacement cannot be avoided. We must live with those additional risks. However, in other cases there may be ways to avoid interventions or extend the intervals between them. If we can, then we are well advised to consider those ideas for implementation as "one-time changes."

We can now see that for any TbM task to be technically feasible, we must have a failure that conforms to one of the failure patterns A, B, and C. In terms of Weibull analysis, it must have a shape parameter (β) that is greater than 1. We must have knowledge of the length of the item's useful life—the point at which its cumulative probability rises above a limit you consider to be tolerable (e.g., 10%). The task frequency is equal to or just below this useful life limit. The replacement or restoration task must also restore the asset to the "as new" condition or as close to it as possible.

Once a technically feasible TbM task has been defined and its frequency identified, we must determine whether it is worth doing. Remember that failures have consequences—safety, environmental, operational, or nonoperational.

If the failure consequences involve safety or the environment, we consider levels of risk in our decision criteria. The execution of the TbM task must reduce the risk from the failure consequences below the tolerable level of risk to safety or the environment that we establish.

For example, let's consider penetration of piping at a bend due to erosion by a slurry. The slurry leak poses both an environmental problem and a safety problem to personnel who come into contact with it. We may want to reduce the occurrences to 1 every 10 years, yet we presently have an MTBF of 1 year. Note that MTBF is the age at which 50% of the failures will have occurred. To reduce incidents ten-fold, we need to replace the piping bend at an interval where the cumulative probability of failure is 1/10 of the 50% probability occurring at the MTBF (i.e., 5%). Using historical failure data, we would carry out Weibull analysis, plot the probability over time, and determine the age at which that 5% cumulative probability of failure is reached.

For operational or nonoperational consequences, decisions are based on costs. The cost of the preventive replacement or restoration must be less than the cost of the failure replacement plus any additional costs incurred if the failure is allowed to occur. These costs include loss of product, loss of production, scrap, rework, additional energy costs, repairs of any secondary damage that might occur, and so on. Normally, these operational costs are quite high relative to the costs of repair, replacement, or restoration. In some cases, it may seem to be a "no-brainer" to make the decision to perform the TbM, but remember to consider how often the TbM task will be performed relative to how often the failure would occur. For instance, in our previous example, if there were no safety or environmental consequences, but the pipe bend were to be replaced at the age where cumulative probability of failure is only 5%, it would be replaced 10 times for every failure that would have occurred at the historical MTBF (50% point). The replacement cost must be considered 10 times for every failure that is avoided.

Any task we choose must meet both sets of criteria—it must be technically feasible, and it must be worth doing the task at the chosen frequency.

PREDICTIVE MAINTENANCE (PDM)

PdM is used to predict when a failure that has already begun will propagate to the point where it manifests as a loss of some functional capability. The failure may be age or usage related, or it may be caused by some random event or condition. In all cases, the failures take some time to develop before loss of function. For PdM to work, the failure mechanism must also provide some indication that it is in progress. That indication is known as a "potential failure condition." Note that with PdM, we are not preventing the failures at all. In every case, the failure mechanism has already begun to propagate, and we are now predicting when we will lose functionality.

Predictive maintenance is often called *condition-based maintenance* or *on-condition maintenance*. It includes two distinct components: a monitoring activity (often called condition monitoring) followed by a restorative action (called *on-condition restoration/repair*, or simply *corrective maintenance*). Keep in mind the difference between this sort of corrective work and corrective work that arises after functional failure has occurred. In this case, we manage when we do the work, creating the opportunity to minimize or mitigate the consequences of the functional failure.

Monitoring is done frequently, while the restoring activity is done once, each time an incipient failure is found. Chapter 9 deals with specific technologies that are used most commonly in carrying out the predictive component of PdM. For our purposes in this chapter, it is sufficient to know that those technologies reveal one failure mechanism or another that is in progress (i.e., a potential failure condition). Generally, the further the failure has progressed, the more severe the indication will be. For example, if we use engine "noise" as a "potential failure" indicator for your car's engine, then the louder that sound becomes, the closer your engine is to failing. The sooner we detect it, the less damage will have occurred before we intervene. If we continue to ignore the noise, then the consequence is engine failure, inconvenience, and a substantial repair bill.

The restorative part of PdM comprises planned and scheduled maintenance. The repair of the failing condition must be done in time to avoid the loss of functionality or the consequences that will arise when functionality is lost (e.g., a bigger repair bill and extended loss of use of the equipment). Timing of that work must be soon after the potential failure is detected. Prompt attention to what we discover through monitoring is critical to success with PdM. Of course, the predictive monitoring activity must give us sufficient warning time to act, so determination of appropriate monitoring intervals is important. For our car engine example, we are able to depend on the most readily available condition monitoring technology we have—the operator's human senses (sound in this case), and we monitor the noise every time we operate the vehicle.

Again, consider the failure patterns presented in Chapter 2. Preventive maintenance (PM) applies only to failure patterns A, B, and C, where we can define the end of useful life. With patterns D, E, and F (predominantly random failure patterns), we have no idea when the failure will occur. However, we can monitor the asset performance, its condition, or both to demonstrate that it is still working well. This works for patterns D, E, and F, and it can also work for A, B, and C as an alternative to TbM, provided they also have potential failure conditions that can be monitored. For example, a bearing that fails randomly provides potential failure warnings through ultrasonic noise, vibrations, particles in its lubricant, and increasing temperatures. A reciprocating compressor valve that fails with usage also gives a warning, since valve temperatures increase as it begins to leak gas back and forth during operation.

At this point in our RCM-R® analysis, we have identified the failure causes. This helps us identify changes to physical, electrical, chemical, or other phenomena that those causes create.

The first criterion for PdM to work is that we must have an identifiable potential failure condition. We must be able to detect that potential failure condition (P). We must then be able to forecast (predict) with some confidence just when the potential failure condition will worsen to the point where we have a loss of function—that is, the functional failure (F). That forecast of time between points P and F is known as the P to F interval (P-F), as shown in Figure 8.2.

Estimating P-F can be challenging, and it is very sensitive to your operating context. Point P varies depending on the potential failure condition you are monitoring and the technology you use. For instance, in a rolling element bearing, we can detect flaws better with some technologies than with others. We might find early crack formation and noise using ultrasonic techniques far sooner than we will find vibrations using accelerometers. We may find vibrations increasing sooner than we detect metal particles in oil samples, and those particles may show up sooner than we are able to notice any significant bearing temperature increases.

The rate of failure propagation (or the rate of deterioration) will vary with your operating conditions. This can move point F. The load at which the system is operating, its environment (e.g., humid, dry, hot, cold), the operating cycle the asset is exposed to (e.g., continuous vs. intermittent vs. batch), the materials being handled by the asset, and so on can all impact

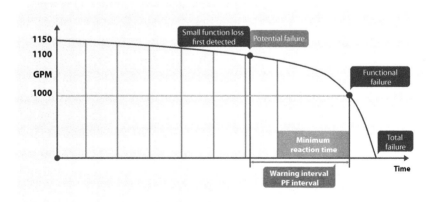

FIGURE 8.2
Condition monitoring and P-F interval.

on how long it takes to deteriorate to a functionally failed state (F) from the time you detect its potential failure condition (P). For example, if you operate at higher loads, then you can expect deterioration to point F to be more rapid. Operational variability can increase stresses. If you are starting and stopping frequently, you can expect more rapid deterioration than if your system operates continuously and at a steady load.

Rarely do we find good maintenance records showing detected potential failures (P) followed by indication of how close we were to failure (F) when we took action. Our maintenance management systems are not designed to record that data, and of course, getting it collected would be an interesting challenge to deal with. Sadly, though, we often find that a potential failure was discovered in our PdM records, then we also discover failures on the same equipment not long afterward. In those cases, the follow-up corrective action didn't take place. It failed in service, and the PdM efforts were wasted!

Your maintenance and operating staff are usually the best source of information for estimating P-F. They work with the assets daily, and they will have a memory (often subconscious) of how long something lasts after some sign of trouble first arises. Once they understand the concept of P-F, many operators and maintainers can give reasonable estimates for it. They must be asked, however, and that is why we include this in the RCM-R® process. Even if they don't consciously think about P-F, their subconscious doesn't miss anything. They will often have a reasonable "gut feel" for P-F. In our experience, it is often the maintainers for a given area of plant or asset class who have the best sense of P-F. Bear in mind that this "gut feel" will be indicative of P-F for whatever potential failure conditions and operating context you have experienced in the past. If all you've ever done is listen to the sounds that equipment makes, then your P will be valid for detecting sounds of trouble. If you switch to more sophisticated technologies, then P will occur sooner and P-F will be longer. If your operating context has changed (becoming either more or less severe), then F will also move.

Once you have an estimate of P-F that your analysis team can accept, you can determine your condition monitoring task interval. The general rule of thumb is that the task interval is half of P-F. If P-F is 2 weeks, then you would monitor the potential failure condition once every week. The next criterion for the task to be technically feasible is that you must be able to act on the findings of your monitoring before you reach point F. If you monitor condition once per week and you find a potential failure, it may be just past the point where you can detect it, or it may be halfway to

F already. It is important to act on that finding before reaching F—in this case, before another week has gone by from the day you found P.

If your P-F interval is quite large, let's say 2 months, then our rule of thumb would suggest you monitor once per month. However, if you can mobilize to repair the defect within a week of finding point P, then you might not need to monitor it as often. For instance, let's say you monitor once a month and find point P as soon as the failure is initiated. You will have a month in which to act. But let's say you only need 1 week to prepare and schedule for this particular repair. If you had acted within that first week after point P, you would have taken the machine out of service up to 5 weeks sooner than you really needed to. Over time, that conservatism will result in more repairs and higher costs. If you just missed P and didn't catch the failure till the next inspection, you would still have a month to act (but you still need only a week). Again, you will take the machine out of service early (3 weeks this time). When determining the timing of your corrective action, consider these factors—in this case, we'd be wise to schedule our repair no later than 3 weeks hence.

Now, consider the planning and preparation time for the job—the minimum reaction time as shown in Figure 8.2. In the figure, the minimum reaction time is large relative to the monitoring interval to ensure we catch the failure with enough of the P-F interval left for action.

In our example, if minimum reaction time is only a week, then you really only need to find point P two weeks before the failure. P-F is still 2 months (8 weeks), but in this case, you could get away with monitoring it every 6 weeks instead of every month. That will reduce our monitoring costs by 50%, and you'll be taking the machine out of service at worst only 1 week before it would have failed. In the worst case, when your monitoring is done just before the failure starts, you will still catch it on your next check at 2 weeks before point F and still have time to act. In those cases of very long P-F, your inspection interval can be lengthened to account for that planning and scheduling lead time before repair is carried out. Caution is advised—you need to be very confident in your P-F estimates, and you need to be very good at work management discipline. If either of those is weak, then you are well advised to stay with the conservative ½ × P-F calculation.

If P-F is very short, then you may not be able to use PdM at all, but if you can find more sensitive technologies for detecting point P (Chapter 9), then you can introduce PdM where it might not have been possible before. In most cases, unless you expect your operating context to change, it won't be practical to do anything to stretch out point F.

Let's say we've found a potential failure condition, we have the necessary monitoring technology, we have a P-F interval that is practical to use (i.e., it produces a reasonable task frequency and leaves enough time to act on any findings). Then, we need to determine whether the task is worth doing.

As with TbM, we consider the consequences—are we concerned with risk or costs? Our task must reduce risks (safety and/or environmental) to a tolerable level, or it must be economically viable compared with the cost of failure. For the cost calculation, remember to consider how often the monitoring will be done in the time it would take to find and proactively avoid the consequences of one failure event occurring at its MTBF.

For example, let's say we have a monitoring task with P-F interval of 1 month. We choose a monitoring interval of 2 weeks. If the failure occurs on average once every 3 years (i.e., MTBF is 3 years), then for each failure event, we "predict" we will have monitored the machine 78 times. We calculate the cost monitoring each time, multiple by 78 and add to that the cost of the repair. We compare that with the cost of repairing a failure plus any other costs associated with the failure (production losses, secondary damage, etc.) and determine which is least expensive.

DETECTIVE MAINTENANCE (FAILURE FINDING) (DM)

DM is done to find failures that have already happened but haven't yet become evident to the operators under normal operating circumstances. It is used where we have redundancy, backups, safety devices, or other protective devices. Those devices are only intended to operate when needed, and at other times they are normally dormant (see Figure 8.3). They are triggered into action by some other event or failure occurring (i.e., the protected function fails). Under normal operating conditions, those triggering events are not occurring, and the devices remain dormant, but they must be available to operate, and they can still fail. Those failures while not in use are "hidden" failures—they can happen at any time before the protected function fails and remain undetected by operators during normal operation, becoming evident only when the protected function fails. In Figure 8.3, Uv represents the unavailability of the protective device after it has failed. For this reason, the operator will see that two failures have occurred—the protected function and the protective device. The use

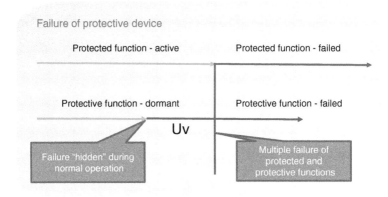

FIGURE 8.3
Multiple failure: Protective device failed before protected function fails.

of protective devices, backups, and redundancy creates the potential for these hidden failures along with a risk of suffering these "multiple failures" and their (usually severe) consequences.

Some care is needed to avoid confusing a hidden failure with a failure of a device that is used intermittently and fails. In the latter case, under normal operating conditions, the device will eventually be called on to work, and its failure will become immediately evident to the operators. It actually fails during "normal operation," and this becomes immediately evident to the operators. An example of this is the brake system in your car. You only use it when you want to slow down or stop—intermittently. If it fails while you are driving, you will know about it as soon as you apply the brakes.

Protective devices, on the other hand, may fail, and their failures may go undetected for a long period of normal operation before the devices are needed. When they are finally needed, the normally operating function that you were protecting will be lost.

In your car's brake system, the loss of brakes is immediately evident; however, failure of your brake lights may not be. In normal operation, you can't see whether the brake lights are working or not, yet you depend on them. If they fail, you may not know about it for some time. Indeed, no circumstance may arise when they are truly needed—for example, no one is tailgating, and other drivers can detect your slowing and stopping in time to avoid hitting you. That can go on for a long time, unless you are unlucky. Then, because you don't have the brake lights, you get hit from behind, or the police ticket you, or someone has the opportunity to inform you that they noticed your brake lights are out. Two of these circumstances are unwanted, so it would be better to avoid them arising as best you can.

These devices are usually designed into systems to protect against unacceptable consequences of failures. They comprise our safety systems and machinery protective devices, to avoid excessive damage; to protect personnel from harm (e.g., guards on rotating equipment); or to catch or avoid situations that could harm personnel (e.g., overspeed trip devices) or the environment (e.g., secondary containment) or have severe production loss penalties (e.g., limiters that prevent overspeeding or overloading). Their function statements often include words like "To act in the event of …" Their statements of effects should also tell us how failures become evident to operators under normal circumstances. For protective devices, these statements often say something like "Under normal conditions this failure mode is not evident, but when needed, we suffer a loss of …"

In many cases, hidden failures occur randomly, so TbM is ineffective. Hidden failures also occur while the devices are dormant, so there are rarely any potential failure conditions we can monitor. Consequently, it is quite common that we cannot prevent or predict the hidden failure. We can only find it after the fact. We do that with failure finding tests, which we refer to as DM.

For DM to work, we must be able to carry out a test of the protective device's functionality at a practical interval without significantly increasing the risk of leaving the device in a failed state. Ideally, the test includes the entire protective circuit and not just parts of it. For example, testing warning lights on a control panel does not tell us that the sensors triggering them are working—we only know the bulbs are functioning. To test the entire circuit, we often need to simulate the alarm condition in a controlled way so we don't increase operational risks. For example, we can manually raise/lower the level in a tank to see that its high-/low-level alarm is tripped.

Like other proactive maintenance (TbM and PdM), our DM tasks must be worth doing. In the case of safety and environmental consequences, we must be able to perform the test at a frequency that will reduce risks associated with the device not working to a tolerable level. In the case of operational or nonoperational consequences, the task must be economical when compared with the costs of leaving the protective device in the failed state.

Protective devices provide protection in the event of failure of some other (protected) function. There are three scenarios we consider:

1. If the protective device fails on its own, there is no consequence. We have lost the protection, and if our testing reveals it is lost, we can correct it without loss of the protected function.
2. If the protected function fails, and the protective device is functional, then we've achieved the whole point of having the protection. We've avoided the consequences associated with the loss of function, and we've saved the day!
3. If the protective device is in a failed state when the protected function fails, we've blown it. We then have two failures at the same time—a multiple failure situation—and we will suffer the consequences of losing the protected function that we were hoping to avoid.

Appendix A presents the mathematics used to derive the formulae for determining the testing intervals (I) for protective devices configured in various ways. The simplest case is determination of the testing interval for a single protective device that operates in parallel to the protected function. The equation that applies to that situation is

$$I = 2 Uv\, Mv$$

where:
I is the inspection interval
Uv is the targeted unavailability of the protective function
Mv is the MTBF of the protective device

There are two important assumptions to consider:

1. The target unavailability is very small (i.e., less than 5%). This is because the derivation assumes a straight-line approximation to an exponential curve near its origin. This assumption is also quite reasonable, because we want our protective devices to be available as much as practicable (i.e., we want low levels of unavailability).
2. The protective device is also assumed to fail randomly. We use the negative exponential equation in our derivation of the formulae. Again, this is reasonable, because most protective devices do fail

randomly. They are normally dormant and not subjected to cyclical usage. They do age, however, and we must watch for protective devices that simply get "stuck" as they do. However, if they do fail with age or usage, then TbM may be more appropriate to use, and testing may not be needed. For example, a valve that is called on to act in a protective circuit may have elastomeric O-rings that can fail with age. It would be better to change those O-rings periodically (thus preventing the valve failure) than to test it and find that it has already failed. Of course, in that case, the rest of the protective circuit may still require testing.

Usually, we find that Uv is not easily measured. Most maintainers and operators don't think in terms of unavailability. However, we show in the Appendix that Uv can be approximated if we know the MTBF of the protected function (Md) and then divide it by the MTBF of the multiple failure (Mm) that we are trying to avoid. Replacing Uv with Md/Mm gives us:

$$I = 2\frac{Md \times Mv}{Mm}$$

We often think in terms of events occurring with probabilities of 1 in 10,000 or 1 in 100,000, and so on. Mm in those cases would be 10,000 and 100,000, respectively.

For example, we might want to reduce the risk of a fatality due a machine failing combined with the likelihood of a person being present at the time to less than 1 in 100,000 in any year. Consider an overhead hoist with an emergency load brake that operates to hold the suspended load in the event of power failure. If the brake fails and no one is underneath it, we avoid the fatality. But, if we estimate there is a 1/1000 chance of a person being under it in any year, then we need to know how often to test the brake so we achieve the 1 in 100,000 chance of injuring someone. Mm is 100,000. We need to know Md and Mv to calculate I.

Md is based on demand, and in this case, it is related to how often someone is likely to be under the load being carried. Since we want a 1/1000 chance of the person being there, Md is 1000 years.

Mv is the MTBF of the brake itself. Let's say that we have found it failed twice in 10 years of inspections. Mv is 10 years/2 failures = 5 years per failure.

Using

$$I = 2\frac{Md \times Mv}{Mm}$$

$I = 2$ (1,000 years × 5 years)/100,000 years = 1/10 year.

We would test the brake every 1.2 months or every 6 weeks.

In the case of risk calculations, the formula will produce a test interval that reduces risk to the level you deem tolerable. But we are not done yet. We must also determine whether it is practical to do the task at the interval we calculate. In our example, testing a crane's load brake once every 6 weeks is practical, so we would accept the task.

We often have protective devices in our production systems for operational reasons—they help us avoid loss of production that might be very costly to our business. In those cases, the formula is modified to consider costs, and for that simple single–protective device configuration, it is:

$$I = \left(\frac{2\,Mv\,Md\,Cff}{Cm}\right)^{1/2}$$

where:

Cff is the cost of the failure finding task

Cm is the cost of the multiple failure we trying to avoid (i.e., repair cost for both protective device and protected device, the cost of lost production, etc., that would be incurred)

In our crane example, let's say there is virtually no risk that a person might be underneath it, but dropping a load would be expensive. The cost of the repairs might be $10,000, but the load it was carrying might be worth $50,000. While the crane is down for repair, production must be curtailed for 8 hours till it is restored, and we lose another $500,000 per 8 hour shift of downtime. Total cost, Cm, is $560,000. The cost of the failure finding task (the brake test) is only $150 each time it is done (Cff). Using the above formula with these costs,

$$I = \left((2 \times 5 \times 1000 \times \$150)/\$560,000\right)^{1/2}$$

$$I = (\$150,000/\$560,000)^{1/2} = 0.04 \text{ year} (15 \text{ days})$$

We would test the brake every 2 weeks in this case. Again, this is a practical interval for testing the brake, so we would accept the task.

There are also a number of other configurations and cases described in Appendix A, together with derivations of formulae for those situations.

Regardless of which situation we are dealing with, if we find that we cannot accept the task at the required frequency for any reason, then we must make some other one-time change to improve the reliability of the device (i.e., increase Mv) or reduce demand (i.e., increase Md) for its use.

RUNNING TO FAILURE (RTF)

RTF is a valid decision for managing failure consequences. If the consequences of failure are minor, then we may be prepared to live with them. If the consequences involve safety or environmental risks, it would be irresponsible to allow run to failure. However, if we are only losing money, then RTF becomes an option. For both operational and nonoperational consequences, RTF is the default action if there is no technically feasible task that we can identify that is worth doing.

Of course, RTF may still be an expensive option, so it may not be an acceptable choice in all cases. In that situation, you are left with no option but to consider one-time changes.

NONRECURRING ACTIONS (ONE-TIME CHANGES)

If we cannot predict, prevent, or detect the failures, and RTF isn't tolerable, then we must change something else. The cause of the failure may or may not be the equipment—it could be operating procedures, repair procedures, or the design of the process in which the equipment operates. It may be that the design is such that it cannot be maintained, or it may not be capable of meeting operational demands. Hidden failures may be undetectable. We need to address whatever shortfall we have through a "one-time change."

Whenever an asset is designed, certain engineering features are built into it, materials are chosen for their strength and resistance to corrosion, tolerances and surface finishes are specified, operating speeds and power consumption are defined and limited, and so on. The asset is then chosen for a given application in your operating context. That operating context may be close to what the designers envisioned, or it may not. It is not uncommon that operators attempt to get more out of their systems than they were originally intended to deliver. Get any of this wrong, and the asset itself may prove to be unreliable in operation or simply incapable of meeting operating expectations.

Reliability is designed into our assets. It is a product of engineering efforts at the design and construction phases of an asset's life cycle. Simpler designs tend to be more reliable, because there are fewer components to fail. They may also be lower cost than more complex designs, also because they have fewer components and simpler assembly. Reliable yet complex designs tend to be more expensive, often because they've used stronger materials, tighter tolerances, higher-quality components, and so on. As you can see, the cheaper design may be better, but that won't always be the case. If high reliability is going to be needed, then it pays to be attentive at the design stage. Inattention can be costly.

If an asset's design falls short of the intended operational requirements, then its performance will be a disappointment. Maintenance costs will likely be high, but maintenance will be able to sustain that built-in capability. However, maintenance cannot raise the capability above what it is designed for without changing the design. Operators contribute to reliability by their care in operating the assets as intended and by keeping demands on the assets within their performance envelopes. If we run the assets harder than they are capable of withstanding, they will eventually fail, and usually sooner rather than later.

One-time changes can make up for any shortfall we identify that results in failures. These changes can include additions or modifications to procedures, processes, skills, and training as well as redesigns of the assets themselves. Once the RCM-R® analysis is completed, these changes must be implemented, often by people or departments other than operations and maintenance. RCM-R® has potential impact across your entire organization. For this reason, it is important to recognize that implementing RCM-R® successfully isn't just a maintenance responsibility.

For instance, if a failure occurs because of operator human error, there is nothing we can do to prevent or predict it, and it certainly won't be hidden. If the cause of that error is a lack of training, then it is the training that needs to be changed. If training is delivered through a training department or human resources, then they too become involved in the implementation of RCM-R® outputs.

If the cause of the failure is some flawed procedure that someone followed, then the procedure must be changed. Likewise, if a procedure or operating instruction is missing, incorrect, unclear, or misleading, it will need to be changed. These sorts of changes can involve engineering, operations, maintenance, and document management groups. They may even require follow-up training to ensure the procedures are understood and used correctly. If maintenance instructions or plans are wrong (e.g., they list the wrong parts), then they must be corrected by maintenance planners. This could also reveal problems in part identification in store rooms and inventory management systems that need to be corrected or the problem will reoccur.

If the machine design is ergonomically unmaintainable or inaccessible, then it must be changed. For example, there may be nothing wrong with a particular machine, but TbM is made all but impossible due to the crowding of pipework, wire trays, structural beams, and other obstructions that render the machine difficult to maintain in situ. An otherwise technically feasible TbM task may be rendered uneconomic.

Some failure causes require technical changes to parts, materials, working fluids, adjustments, speeds, and so on. System design may be prone to failure. We may find that the machine is wrong for the functional performance being asked of it, or that it was a poor choice in the operating context at a particular site. Perhaps the operating context changed in the time period between design and commissioning of a new system. Even reusing a design in a different location can result in a change of operating context. For example, a cooling water pump that works well with fresh water may suffer more corrosion failures at an identical sister plant situated by a body of salt water. That same plant may suffer more blockage due to marine growth inside heat exchangers.

Equipment can become less suitable as time progresses and the operating context evolves away from the original design concept. For example, an electric utility may find that time-of-use pricing spreads load demand out over a longer time period and reduces peak demand, but it also leaves less cool-down time for transformers in its network. A transformer that could easily handle the peaks and average load is now seeing increased average loads (due to the proliferation of consumer electronics) and shorter cooling periods—its transformers are now failing more frequently.

DESIGN CHANGE OBJECTIVES

For safety or environmental consequences, our design changes are intended to reduce those consequences to tolerable levels or eliminate them altogether. We could replace the design with one that is more "fail safe." We might consider improving the reliability of a protected function so it is less likely to be needed. We can enhance protection by adding containment devices or boundaries to the design. If the failure is hidden, we could consider making it evident or adding a protective function.

If a failure is hidden but not testable, we may consider changing it to a design that can be tested. We could add redundant protective devices to increase the availability of the protection or make the device(s) more reliable.

If any scheduled maintenance is technically feasible, but it isn't worth doing, then we can change the design to make the work more accessible and less costly. We could also improve the reliability of the function so that it fails less often, requiring fewer maintenance interventions. That won't change the cost of the maintenance activity, but it will reduce how often it is done and may make it more cost effective.

If our failure consequences are operational or nonoperational, we can change the design to reduce failure frequency (improve the reliability of components), or we can eliminate the consequences by adding redundancy or backup capacity. Note that this adds to maintenance overall, but the investment of capital may still be justified.

TIMING

If failures are a result of original design flaws, then they can be dealt with easily if our analysis is being done during the design process. One of the strengths of RCM is its value at the design stage in identifying just this sort of problem early, while it is still easy to change it.

Doing meaningful analysis implies the need to involve maintainers in the design stage and to carry out this sort of analysis on the design as it evolves. This is how RCM has been applied in aircraft and military systems, and it has been hugely successful there. There is an old carpenters' saying, "measure twice and cut once." It is costly to waste materials and time to do the work over. That saying speaks to the value of planning ahead and taking care to get it right. Unfortunately, few companies make this investment, despite its high potential for payback. They are saving on initial investment but paying for it later—fooling themselves with false economy.

If we carry out RCM-R® during the operational phase of the asset life cycle, then these problems can still arise, and we will have no choice but to deal with them. Unfortunately, any change to the physical asset, its configuration, or the systems within which it is built can be a very expensive proposition. Changing a plant or process that is already built is costly, not to mention disruptive of revenue streams. RCM-R® will identify these situations, but only after investigating all other possible solutions through maintenance.

DESCRIBING ONE-TIME CHANGES

When an analysis team identifies the need for a one-time change, it must keep in mind that it is only stating a requirement. The writing or rewriting of procedures, development of training, identification of trainers, and redesign of the asset are all activities that are done after the RCM-R® analysis is completed. The analysis team should describe the need for the change and what it is intended to accomplish. There is no need to define exactly what the change will be, although if they have ideas, then they are encouraged to include them. Ultimately, however, the decision on the details of those changes will lie in the hands of those who write procedures, provide training, and design your facilities.

CHAPTER SUMMARY

This chapter has presented the various options that are available for managing the consequences of failures. Recurring preventive, predictive, or detective tasks, whether they are performed by maintainers or operators, deal with the failures and their causes with the aim of reducing consequences to levels we can tolerate. Each of these tasks has a task interval (frequency) that is determined on the basis of sound technical criteria relevant to the operating context in which the failure occurs.

PM is used for failures that are triggered by age or usage factors to restore or replace functionality before it fails. They actually prevent the failure from occurring.

PdM is used to identify failures that have begun but not yet progressed to a loss of functionality. It works for failures that cannot be prevented because they are random in nature, and in some cases, it works on aging failures, where they provide early signs of deterioration. Several of the more common and popular PM techniques are described in more detail in the next chapter.

DM is used to find failures that have already occurred in protective devices that may otherwise go undetected during normal operation of the assets.

In all cases where we have operational or nonoperational consequences, if we can't find suitable tasks that are both technically feasible and worth doing from a cost perspective, then we can always default to running the item to failure. Sometimes that is not an attractive option, and we may still want to do more anyway.

Similarly, in situations where we have safety or environmental consequences, and we can't find suitable tasks that are technically feasible and worth doing from a risk perspective, then we must find some other alternative than RTF. Those alternatives to a RTF strategy include a variety of one-time changes to the design of the item, the design of its protective system, training in the event of human error, or procedures that might be missing or flawed and therefore lead to mistakes being made.

All of these are strategies to minimize or eliminate the consequences of failures. Now, we look at specific technologies for PM (Chapter 9) and how we go about tying this all together in a strategy selection process (Chapter 10).

9

Condition-Based Maintenance Techniques

RCM-R® produces decisions, many of which require the use of condition-based maintenance (CBM). Going back to the basics as described in Chapter 2, it is evident that the majority of failure modes will be random in nature. Because there is no relationship between age or service life of the asset and when it will fail randomly, preventive techniques (restoration and replacement) will be of no value in these cases. Many failures, whether they develop with age or usage, or whether they are random in nature, do provide some early warning of their development. Once the failure mechanism has begun, it will propagate and eventually lead to the failure mode event. In doing so, the item that is failing becomes "weaker"—less resistant to failure. In fact, it is already failing. We are well advised to detect these early signs that reveal these incipient failures before they propagate to the failed state. CBM is intended to do just that—give us an early warning of the failures as they are propagating. We can rely on our human senses for some of this work (e.g., we can hear parts that are rubbing; we can feel elevated temperatures and vibration levels), but we can also employ an array of technologies that can detect these "signals" much sooner than our human senses. One of the big advantages of CBM, in addition to giving us advanced warning of failures, is that much of it is nonintrusive—CBM can often be carried out without the need to disturb equipment operation. Of course, the earlier we can detect failures in progress, the earlier we can take steps to mitigate the consequences of those failures. We have time to arrange for parts, tools, alternative means of production, and so on. This chapter explores the more widely used CBM techniques: vibration analysis, infrared thermography, precision lubrication and oil analysis, ultrasonic surveys, and nondestructive testing.

The authors are thankful to several colleagues who have contributed to this chapter. They are named at the start of their contributions.

VIBRATION ANALYSIS

Jesús R. Sifonte

In our context, vibration is defined as a pulsating motion a machine experiences from an origin location at rest. Machine supports react to internally applied forces with vibration as the machine operates. Machine vibration analysis is the study of the behavior of rotating machinery for detecting faults based on monitoring and trending vibration signals produced by its components. All rotating machinery, whether it is in good or bad condition, will vibrate. A vibration analysis can help us determine whether the machinery vibration is normal or not. It may even detect many faults early enough to plan repairs at a convenient moment, avoiding costly plant shutdowns. The purpose of vibration analysis within the predictive maintenance context is to determine the machine's health while it is operating.

There are misconceptions about machinery vibration. For instance, people often think that machines vibrating more are in worse condition than others with less vibration. That is not always the case.

The most basic vibration principle establishes that vibration amplitude or intensity is proportional to the forces causing it, and at the same time, it is inversely proportional to the dynamic resistance the machine offers to the applied forces. Figure 9.1 illustrates that motion is produced by the internal forces coming from the machine's rotating components. The machine's weight, together with its foundation base, offers opposition to that motion, which is called *dynamic resistance*. The total travel that the machine bearing housing is able to attain is the result of the opposition of internal forces trying to move the housing and the dynamic resistance offered by the machine and its base structure.

Vibration Parameters and Units

Single degree of freedom systems (i.e., those that can move in only one direction) undergo the simplest possible vibratory motion. The resulting vibration time waveform in a single direction (let's say in the radial

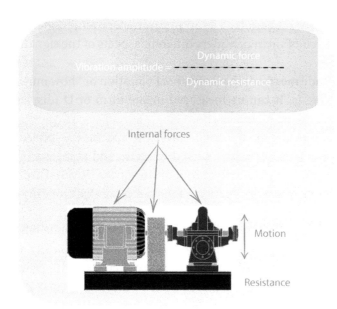

FIGURE 9.1
Vibration amplitude concept.

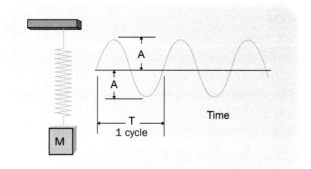

FIGURE 9.2
Simple harmonic motion.

direction) of an ideal rotor or thin disk while it is rotating is called *simple harmonic motion*. Here, the rotor is represented simply as a weight on a spring. Its movement can be represented as a sine wave, as plotted in Figure 9.2. Note that the amplitude A in the figure is a representation of the travel the rotor surface experiences as measured from its rest position to the maximum displacement in either the up or the down position. T, called the *period* of the sine wave, equals the time it takes the rotor to complete a single up and down cycle.

There are three measurable vibration parameters: amplitude, frequency, and phase. Each of them considers different aspects of the signal:

- Amplitude measures the intensity of vibration or "how much" vibration there is. It can be measured in the form of D (displacement), V (velocity), or A (acceleration).
- D is a measure of the total travel the machine experiences (i.e., $2 \times A$ in Figure 9.2). It is related to stress and is measured in mils peak to peak or millimeters root mean square (RMS). Displacement is used in low-speed situations, for example, a shaft rotating at under 600 RPM.
- V is the velocity the machine surface undertakes while vibrating. It relates to the fatigue the surface endures, and it is typically measured as inches per second (ips) peak, velocity decibels (VdB) peak, or millimeters per second peak. It is used in the 600–10,000 RPM shaft speed range.
- A is the acceleration experienced while the machine is in motion. It is related to force. It is measured in Gs (gravitational acceleration units), and it is used in high-speed shaft measurements over 10,000 RPM.
- Frequency measures how "many times per second" the signal repeats itself. It is used to determine particular sources of vibration within complex (nonsimple) harmonic motion. For example, a $1 \times$ RPM vibrational frequency often indicates machine imbalance, while $2 \times$ RPM often indicates parallel misalignment between components. It is measured in cycles per minute (CPM), Hertz, or orders (multiples of the machine driver component's rotational speed).
- Phase measures "how" the motion being experienced at a particular spot or direction relates to the motion in other parts of the machine. Phase is useful to help determine multiple sources of vibration—for example, two unbalanced spots on a rotor each produce a vibration but at different locations; combining the two gives us a single signal representing the addition of the two forces at a third, phase-shifted location. Since we can only detect it at fixed locations (i.e., bearing housings), we will see slightly different vibrations at each housing. We can use this to help us in correcting the balance of long rotors, where the exact location of any imbalance is all but impossible to detect. We also use this information, in a simple form, to balance

wheels and tires on cars. It is measured in degrees or radians from a reference location.

Vibration Analyses

There are various types and levels of vibration analyses requiring varying degrees of expertise to apply correctly. Overall vibration analyses, for example, only require an overall vibration meter (usually a pen-type meter) and little training, while spectral analyses need sophisticated and more expensive Fast Fourier Transform (FFT) analyzers, demanding a lot of expertise from the analysts to interpret the outputs.

The more sophisticated spectral analysis is capable of pinpointing vibration sources (machine internal components) precisely, while the overall analysis can only measure the sum of all the vibration produced by the machine in the 10–1000 Hz range. It is often related to the overall condition of the machine. Overall vibration tells you there is a problem; spectral analysis tells you what the problem is.

A vibration time waveform is converted into a spectrum through the FFT process. The vibration spectrum is an amplitude versus frequency graph, making possible the identification of complex machinery vibration, as shown in Figure 9.3. Spectral analysis entails taking vibration measurements on rotating machinery bearing caps, relating the spectral peaks to machinery component rotational speeds and passing element

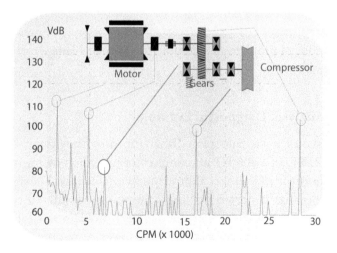

FIGURE 9.3
Vibration spectrum components.

events, trending the signals over time, and producing machine condition diagnostics and recommendations.

For example, $1 \times$ RPM vibrations are related to shaft speed—possibly imbalances. Vibrations can also appear at gear meshing frequencies, bearing ball passing frequencies, $2 \times$ and higher multiples of shaft speed, rotor natural frequencies, and so on. Each of these frequencies identifies a source of vibration; the amplitude at each frequency gives us a sense of how that vibration compares with "normal" levels from baseline readings.

Machine Condition Diagnosis with Vibration Analyses

Spectral vibration analysis is capable of detecting many faults at fairly early stages in their development and without disrupting machine operation. This makes vibration analysis a truly valuable CBM technique. It is commonly used to avoid unnecessary time-based repairs or changes. It helps us to avoid reducing the useful life of components that otherwise might be changed preventively and to avoid inducing premature failures. Some of the most common conditions diagnosed with spectral vibration analysis are

- Rotor imbalance and eccentricity
- Shaft misalignment and bent shafts
- Mechanical looseness
- Antifriction and journal bearing wear
- Electrically induced faults
- Gear problems
- Flow-induced problems (cavitation, turbulence, blade wear)
- Belt wear

Vibration Analysis Diagnostic Example

Figure 9.4 shows a cascade plot displaying how the Rotor Bar Pass Frequency (RBPF) of a 400 HP air compressor induction motor increased over time. Unfortunately, it was almost run to total failure. The motor was taken apart because it started making a humming sound and two of its rotor bars were broken. Note that the incipient problem was found with a 0.06 in/s Peak (PK) on October 10, 2003, some 13 months prior to the much higher 0.33 in/s Peak (PK) reading on December 7, 2004. Total failure was avoided because the analyst observed the RBPF component single

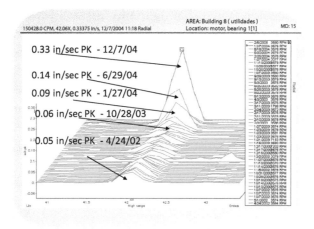

FIGURE 9.4
Diagnosed broken rotor bar vibration spectra.

as it continued to increase, eventually reaching an unacceptably high level of vibration, indicating the possibility of severe damage. See the January 27, 2004 and June 29, 2004 measurements, showing some 80% and 180% increase in RBPF vibration amplitude over the base 0.05 in/s vibration noted on April 24, 2002. This case study shows how spectral vibration can be used to detect even less common rotating machinery problems, such as a loose rotor bar in a motor. Trending over time is what really makes vibration (and any other CBM technique) a highly valuable tool for organizations to attain their business goals, helping to maximize uptime by stretching the working life of components as far as practicable without allowing catastrophic failure.

INFRARED THERMOGRAPHY

Wayne Ruddock

Introduction and History

In 1678, Christiaan Huygens put forward his theory that visible light was in the form of waves. On February 11, 1800, William Herschel discovered that there were invisible "dark heat waves" off the red end of the visible spectrum, which he called infrared waves. In 1865, James

Clerk Maxwell demonstrated that light and infrared waves were both forms of what we now know as electromagnetic waves. Infrared thermography is the viewing, analyzing, and saving of an infrared image with an infrared imaging device. If there is no image, then it is not thermography. There are currently two common infrared devices used in predictive maintenance: infrared cameras and infrared radiometers. Infrared radiometers, commonly known as *spot meters*, are not infrared thermography devices, as there is no image or picture produced. These handheld "pistols" simply measure the radiated energy of a specified circular area, and the built-in microprocessor calculates the temperature and displays that value as a digital readout on a display screen on the back of the gun.

Basic Infrared Theory

All objects that exist above absolute zero ($-273°C$, $-459°F$) give off invisible infrared radiation whenever a suitable medium is present. That medium can be a gas or a vacuum. Infrared radiation does not travel through most solids and liquids. The first law of infrared thermography states that *infrared instruments see ONLY the radiated energy from the first 1/1000 of an inch of the surface of most solids and liquids. They do not see temperature, and they do not measure temperature.* This applies to all instruments that operate in the wavelength band between 2 and 14 microns, or millionths of a meter. Most infrared devices today have some type of computer system built in to interpret the radiated energy they detect. Once the infrared device measures the radiated energy, the computer will calculate the temperature using the Stefan–Boltzmann relationship.

We know that unfortunately, objects in reality do not give off infrared radiation at the same rate, and this complicates our task. In theory, we would expect the following:

1. Infrared energy is emitted from the surface of all objects above absolute zero, due to the fact that at temperatures above absolute zero, the object has energy, which causes the molecules on the surface of the substance to vibrate.
2. Temperature can be defined as a measure of the average kinetic vibrational energy of the molecules that the temperature measurement is relating to.

3. Theoretically, two objects at the same temperature would give off the same amount of infrared radiation and would look the same in the infrared camera. An object at a higher temperature would give off more radiation than an object at a lower temperature and look hotter than the cooler object in the camera.

However, since objects don't emit infrared energy at the same rates, we don't see this happening in the real world. An object with a temperature of 200°C can look to an infrared camera much cooler than an object at 75°C. On the other hand, an object at 50°C could appear to be much hotter than an object at 200°C. In the real world, objects radiate energy at different rates.

This fact made it almost impossible to determine the behavior of infrared radiation till, in the 1860s, a man named Gustav Kirchoff coined the term *blackbody*. A blackbody is a theoretical object that would emit a maximum amount of energy at any temperature and in any wavelength. It would not only be a perfect emitter, but it would also be a perfect absorber. Unfortunately, in the world in which we live, there is no such thing as a perfect blackbody. However, this concept did allow others to work on the principles of infrared radiation and how it behaved. In 1879, Joseph Stefan by experiment, and again in 1884, Ludwig Boltzmann by theory, determined the relationship between radiated energy and temperature. The relationship is defined by the Stefan–Boltzmann formula:

$$Q = 5.6703 \times 10^{-8} \times T^4 \, (\text{kelvins})$$

where:

Q	= the total amount of radiated energy
5.6703×10^{-8}	= the Stefan–Boltzmann constant
T^4 (kelvins)	= the temperature of the object in kelvins raised to the fourth power

This formula provided a method to calculate the temperature of an object when the amount of blackbody radiation was known. In the RCM predictive maintenance world, *there are no objects that are perfect emitters*. The radiated infrared energy that comes off the surface of an object is a combination of *emitted energy* due to the temperature of an object and the *reflected energy*, which comes from the background and is reflected off the surface of our object of interest. To calculate the correct surface temperature of an object with today's infrared cameras, the technician must account for the emissivity of the object of interest as well as the amount of background reflected energy. If the technician does not manually input the correct value for emissivity and background reflected energy into the computer built into today's infrared cameras and spot radiometers, then every temperature displayed on these devices will be incorrect.

Emissivity is the rate at which an object emits energy compared with that of a blackbody at a given temperature and in a given wavelength. Emissivity is determined by five main object characteristics:

1. The material of the object
2. The surface condition of the first 1/1000 of an inch
3. The temperature of the object
4. The wavelength of the device used to measure the energy
5. The geometry of the area viewed

Applications of Infrared Thermography in Predictive Maintenance (PdM)

There are four main areas where infrared thermography is valuable in any PdM program:

1. Electrical equipment CBM
2. Mechanical equipment CBM
3. Process equipment CBM
4. Facility CBM

The first law of thermodynamics states that energy cannot be created or destroyed in a closed system, but it can be changed from one form to another. This is the basis of the second law of infrared thermography, which is valid for all applications using an infrared camera for predictive maintenance: "Without a driving force to produce a radiated energy difference on the surface of an object, infrared thermography will not work." Simply, if there is no energy in a system, there will be no energy coming out of that system to detect.

Electrical Inspections

The power consumed in an electrical system, whether or not it is used for profitable work, can be defined by a simple formula:

$$\text{Power consumed} = I^2 (\text{current}) \times \text{resistance}$$

Of course, we know that this energy is not really consumed (first law of thermodynamics); it is changed into other forms, including mechanical work (what we usually want) and heat or thermal energy. It is this thermal energy that causes the change in radiated energy that we can observe with an infrared camera. For success with the camera, the infrared thermographer must be cognizant of two foundation principles:

1. An electrical system should be under at least 60% of its normal full load (100% is better) before a thermographic inspection can be performed.
2. All electrical problems detected by infrared are caused by either resistance or load (current).

With higher electrical resistance, it is always hotter at the point of higher resistance. The radiated energy level fades to normal as we get further away from the source (second law of thermodynamics). When using infrared in electrical inspections, we need to understand that higher resistance is not just caused by a loose connection. It can be caused by five different conditions:

1. Overtightened connections, often due to a misconceived preventive maintenance program
2. Oxidized or dirty connection
3. Improper components or materials
4. Poor workmanship
5. Looseness

It is good practice to avoid writing "loose" on a report, as this will result in the component simply being tightened. In many cases, this will only make the problem worse rather than being a solution. It should be described as a bad connection and should be investigated to determine the real problem so a proper repair can be performed and the problem rectified. Figure 9.5 illustrates a bad electrical connection on the middle phase wire at the junction box.

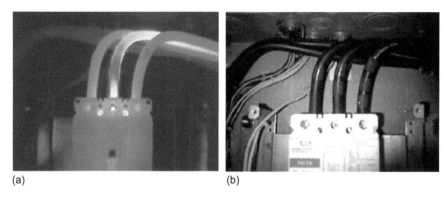

(a) (b)

FIGURE 9.5
Typical radiated energy pattern of a bad connection. (a) Infrared image, (b) what we see with our eyes.

FIGURE 9.6
Infrared signal induced along full length of two conductors in a cable tray.

A current or loading problem has a different pattern from that of a resistance problem. A loading problem shows higher radiated energy for the entire length of the conductor as long as the resilstance is the same, as shown in Figure 9.6. Figure 9.7 shows why—the current induces a magnetic field. Consequently, detecting this will not always indicate a problem. It could be overloaded, or it simply could be an unavoidable imbalance on a three-phase system. Before calling a difference in load a problem, the thermographer must determine the current on the system in question.

A third condition that can be detected by infrared thermography is induced heating. This occurs when a ferrous metal such as an iron bolt is located where it interferes with the magnetic fields produced by electrical currents. It often occurs with the improper installation of high-voltage conductors. Some consider it a problem, and others do not. Figure 9.8 illustrates the induced heating of the two bolts, pointed out using arrows.

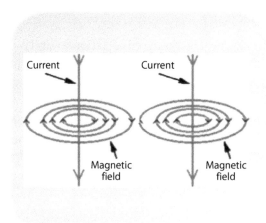

FIGURE 9.7
Magnetic field induced by current. If current is uneven, the magnetic field will be different, and so will the radiated energy.

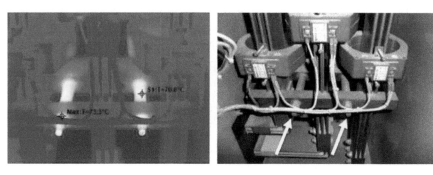

FIGURE 9.8
Induced heating of two bolts used in the assembly, but not a part of the electrical circuit.

The benefits of using infrared thermography for electrical inspections are that it

1. Helps to identify electrical problems, sometimes years before catastrophic failure
2. Increases safety by identifying electrical hazards and areas of potential electrical fires
3. Identifies potential electrical breakdowns, preventing production downtime
4. Provides us with a proper database record that enables our inspection program to mature from reactive to proactive

Mechanical Inspections

Most mechanical equipment is composed of stationary and moving parts. In a perfect system, all of the energy supplied to the machinery (say, by an electric motor) would result in profitable work done by the equipment. However, in reality, a certain amount of energy is transformed by friction or stress into thermal energy. This increases the temperature of the components and usually changes the radiated energy patterns. As mechanical equipment begins to fail or operate in an undesirable fashion, there will usually be a change in the thermal pattern of the equipment as compared with its normal thermal pattern when operating properly. Infrared thermography is one of a number of tools to aid in identifying failing mechanical equipment before it fails catastrophically, causing loss of production or revenues or other consequences.

With mechanical inspections, the thermographer must understand the equipment being inspected and know the normal thermal pattern of that equipment. As with electrical infrared inspections, the equipment should be under normal full load. The greater the mass of the equipment, the longer it should operate before being inspected, so that it is fully "warmed up." The energy must make its way to the surface before being inspected.

Note, too, that not all mechanical equipment is suitable for an infrared predictive maintenance inspection program. Equipment should be chosen using the principles of a document such as ISO 13379 and related ISO standards.

The main principles to consider in analyzing mechanical equipment are

1. Thermal patterns and signatures
2. Actual temperature of the component
3. Ambient temperature where the equipment is being operated
4. Rise above ambient
5. Trending of temperature over time
6. Comparison with similar equipment operating under the same load and conditions

Under the proper conditions, infrared can be used to inspect mechanical equipment such as motors, pumps, bearings, shafts, gears, pulleys, conveyors, fans, drives, compressors, condensers, generators, couplings, and belt drives, as well as large rotating machines comprising many components, such as paper machines and rotary lime kilns, to name a few. Several examples of excessive mechanical heating due to equipment problems are shown in Figures 9.9 through 9.11. Figure 9.12 shows overheating in an electrical motor.

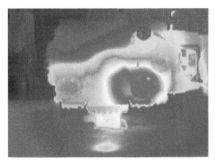 

FIGURE 9.9
Gas compressor with abnormal signature indicating a faulty exhaust valve.

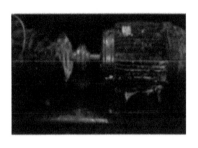 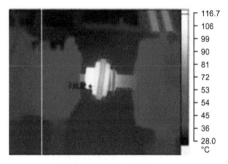

FIGURE 9.10
Misaligned coupling between pump and drive motor.

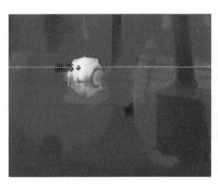 

FIGURE 9.11
Bearing problem.

The benefits of an infrared mechanical inspection program are:

1. It gives us the fastest and least expensive method for identifying mechanical problems, making it very handy as a mass screening tool.

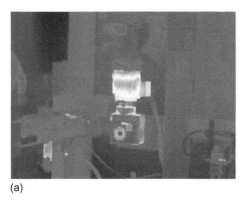 

(a)   (b)

FIGURE 9.12

(a) Normal motor pattern, (b) abnormal (overheating) motor pattern. These are comparative images taken of the same equipment operating at the same load, in the same conditions, and with the same infrared camera settings.

2. It identifies mechanical problems long before catastrophic failure, avoiding lost production, higher repair costs, or other consequences.
3. It lowers repair costs by avoiding collateral damage to associated equipment.
4. It provides us with a proper database of visual records to move our program from reactive to proactive.

Process Inspections

Any time we have a heated or cooled product or process, there is an opportunity to find valuable information using infrared thermography. Process inspections with infrared thermography can cover a wide variety of situations. In insulated systems, a failure in the insulation will appear as a temperature difference on the outside of the equipment, thus producing a difference in the radiated energy. In the case of furnaces and ovens, where the thermal conductivity of the refractory is known, the refractory thickness can often be estimated using thermal surface maps and temperatures.

In many uninsulated pipes, a blockage or flow restriction can result in a variation in surface temperatures. In the case of a warm fluid, a blockage or restriction will appear as a cool spot. In the case of a relatively cool fluid, a blockage or restriction will appear as a warm area.

Failed valves and traps can be located by looking at intake versus outlet temperatures. In most cases, there should be a substantial difference.

With steam traps, it is often necessary to wait till they cycle and compare the inlet/outlet temperatures while cycling with those taken under normal (closed) conditions. If the differences in temperature values are small when compared for the two states, the indications are that you have a failed trap.

Knowing that different materials have different thermal capacities, we can also determine things such as fluid and sludge levels in tanks. Infrared thermography can also be an indicator of the contents of a tank if the materials have different thermal absorption rates. As always with infrared thermography, the process should be in operation to produce a driving force for the thermal energy.

Facility Inspections

In some climates, building inspections are seen as a valuable role for infrared thermography. Where the climate can be either very cold or hot, insulation is usually used in buildings to reduce heat loss or gain while maintaining a large temperature difference between the desired inside temperature and the outside temperature, $10°C–30°C$. For example, in Canada, where winters are long and cold, insulation is used to inhibit interior heat loss from buildings to the cold outside temperatures.

Thermography can be used to identify

1. Missing insulation
2. Damaged insulation (e.g., wet)
3. Water ingress
4. Air infiltration
5. Air exfiltration
6. Excessive thermal bridging (e.g., window frames)
7. Leakage in hot water heating systems
8. Flat roof leaks

Building inspections need to be carried out when there is at least a $10°C$ temperature difference between the inside and the outside of the building. Different types of faults are identified by their specific thermal patterns.

It is also best to inspect buildings at night to negate the effects of thermal gain. Inspections are done from the outside first, giving a general view of the exterior surfaces. Any suspect areas are then investigated with an inspection of the inside surfaces.

Infrared thermography can quickly and accurately identify any area on a conventional built-up flat roof where the water is leaking into the roof system. In this case, there is a large difference in the thermal capacitance of wet insulation as compared with dry insulation, so wet areas will appear to be warmer, as shown in Figure 9.13. It can also be performed in the winter in the northern climates, when the difference in resistance between wet and dry insulation can be evaluated as illustrated in Figure 9.14. In general, it is a summer application. The sun heats up the entire roof area on a hot, relatively calm day. When the sun goes down, the dry roof area loses its energy to the cool night sky at a relatively fast rate. Due to the high specific heat of the wet insulation, the wet areas, on the other hand,

FIGURE 9.13
Refractory problem in a process vessel.

FIGURE 9.14
Area of wet insulation on flat roof.

can hold their energy up to 6 h after the dry insulation has cooled off. It is relatively easy to identify the area of the roof that needs repair using infrared thermography.

PRECISION LUBRICATION AND OIL ANALYSIS

Mark Barnes

Performed correctly, the outcome of any RCM process should result in an optimized series of preventive (time-based) tasks, predictive (condition-based) tasks, and one-time changes. When executed according to the prescribed schedule, these mitigate the causes of potential failure modes. The resultant program should prescribe the most effective condition monitoring tool or technology that will permit early enough warning of an impending problem so that corrective action can be taken before catastrophic failure occurs. And while the number of different failure modes vary widely based on asset type and working environment, for rotating and reciprocating assets, many of the prescribed maintenance activities that come out of the RCM-R® analysis involve basic CBM tasks and activities to either eliminate or identify lubrication-related failures. As such, it stands to reason that any effective RCM-based maintenance strategy should include rigorous control over lubrication practices, while deploying oil analysis to identify lubrication-related problems in a timely fashion.

Identifying Lubrication-Related Failure Modes

The number and type of lubrication-related failure modes that need to be addressed and the appropriate CBM technique to deploy will vary greatly by asset type. For example, the ways in which a high-pressure servo-controlled hydraulic system might fail and should be monitored will be vastly different from those of a slow-turning splash-lubricated gear reducer, which, in turn, will vary widely from a turbo compressor train in a refinery. However, for all oil- or grease-lubricated equipment, there are 10 basic failures that should be addressed in the RCM analysis. These are outlined in Table 9.1, together with the appropriate preventive, predictive, or proactive actions to address each one.

TABLE 9.1

Lubrication-Related Failure Modes and Their Management Strategies

Failure Mode	Appropriate Preventive Maintenance Task	Predictive Maintenance Strategy	One-Time Changes/Enablers
Lack of lubrication	Route-based lubrication rounds that include all lubricated assets	Visual oil level checks, ultrasonic greased bearing inspections, high-frequency vibration analysis	Proper level gauges installed and marked with high/low running level, UV leak testing
Too much lubricant (overlubrication)	Prescriptive task details that include the correct amount of lubricant to apply	Visual oil level checks, ultrasonic greased bearing inspections, high-frequency vibration analysis	Properly engineered PM tasks
Wrong lubricant selected	None	None	Lubricant survey completed
Wrong lubricant added	None	Oil analysis (viscosity, additives, etc.)	Appropriate tagging of lubricant application and transfer points
Lubricant contaminated with moisture	Periodic or on-condition off-line filtration	Oil analysis (water), visual inspection	Proper breathers installed. Proper seal management
Lubricant contaminated with particles	Periodic or on-condition off-line filtration	Oil analysis (particles), visual inspection, patch testing	Proper breathers, seals, and filtration
Lubricant degraded	Timely oil changes	Oil analysis (viscosity, acid number, additives, etc.)	None
Lubricant too hot/cold	None	Basic temperature checks, thermography	None
Additives depleted	Timely oil changes	Oil analysis	None
Lubricant contains foam/air entrainment	None	Visual inspection, oil analysis	Proper system design, clean dry oil, correct oil level

Lack of Lubrication

Perhaps the most basic CBM activity for a lubricated component is to ensure that the correct amount of oil or grease has been added. For oil-lubricated assets, this is usually fairly straightforward; we simply pour or pump the correct oil into the oil sump or reservoir till the level matches the "full" mark provided by the original equipment manufacturer (OEM) and perform periodic visual checks to ensure that the level remains correct. However, in many wet-sump applications, such as splash-lubricated gears and some pumps, the OEM level is provided using a level plug or dipstick, both of which only really provide a proper level when the machine is shut down. In this case, it is wise to modify the component to include an external liquid level gauge so that the correct oil level can be read, whether the machine is running or shut down (Figure 9.15). Basic inspections should be included as part of the CBM program to ensure that the oil level is correct.

For grease-lubricated assets, the issue is not quite so straightforward. Many greased components do not come tagged with the correct amount of grease, and even if they do, ensuring that grease is getting to the component—whether through manual or automatic application—is critical. For manually greased application, fill lines should be periodically checked to make sure they are not plugged. For automatic application, the lubrication system should be checked for proper operation, including pump actuation, line blockage, and actuator operation.

FIGURE 9.15
An external level gauge showing high/low running points is an excellent visual CBM tool.

Too Much Lubricant (Overlubrication)

Too much oil in a system can result in oil leakage and contribute to the formation of foam and aerated oil, making oil level checks an important inspection tool. However, for grease-lubricated assets, the problem is much worse. For high-speed greased bearings such as motors and fans, too much grease can cause excess heat to build up, creating increased friction within the moving surfaces. For this reason, proactively identifying the correct quantity of grease to apply at the right frequency is critical. Identifying the correct quantity and frequency to grease is fairly straightforward, requiring just a few basic inputs, such as bearing type (ball bearing, tapered roller bearing, etc.), bearing dimensions, shaft speed, and load. Despite this, problems associated with overgreasing are widespread in plants that have yet to proactively address common lubrication-related failure modes. More progressive organizations have started using high-frequency vibration analysis and/or ultrasonic monitoring to ensure optimum grease volumes with excellent success.

Wrong Lubricant Selected

Lubricant selection is based on load, speed, and operating context. At a basic level, selecting the correct lubricant requires that the correct base oil type (mineral or synthetic), base oil viscosity, and additive package be chosen based on the application. For oil-lubricated assets, most OEMs provide lubricant recommendations, which are always a good starting point. However, they should always be reviewed, and where necessary, changes should be made based on operating conditions (particularly high or low temperatures) as well as load. With the exception of electric motors, few, if any, greased assets come with OEM recommendations. The reason for this is fairly straightforward. When buying bearings, an OEM is typically unaware of the operating load, speed, and application, and as such, is unable to make an accurate lubricant selection. Most misapplication of lubricants in plants involves using the wrong grease. To avoid this, careful consideration of operating conditions, including load and speed, should be made to ensure the correct viscosity of base oil contained within the grease. In addition, the correct grease thickener type (lithium, polyurea, calcium complex, etc.) should be selected based on operating conditions, while avoiding mixing different thickener types to prevent chemical incompatibility. More advanced oil analysis tests such as ferrographic analysis can help determine whether the selection of an incorrect lubricant is causing failure by identifying the morphological properties of any wear debris.

Wrong Lubricant Added

Making sure the correct oil or grease is added is critical to preventing equipment failure through incorrect lubrication specifications or chemical incompatibility. To do this, any lubricant application or transfer point should be tagged with the type of lubricant in use. Best practice is to use a color- and shape-coded tag such as that shown in Figure 9.16 to provide easy and clear identification. Wherever possible, the tag should avoid the use of the brand name of the lubricant to make retagging unnecessary in the event of a change of vendor or a branding name change by the lubricant vendor. Comparing base oil viscosity and additive elemental content using oil analysis is an excellent way to identify whether the wrong lubricant has been added.

Lubricant Contaminated with Moisture

Water in oil can result in rust and corrosion within the oil sump, while moisture in the load zone of a bearing or gear mesh can result in loss of oil film or cavitation. Water can also compromise additive health, while in some ester-type base oils, water can cause base oil degradation. The

FIGURE 9.16
Simple color- and shape-coded tags help to avoid accidental mixing of lubricants.

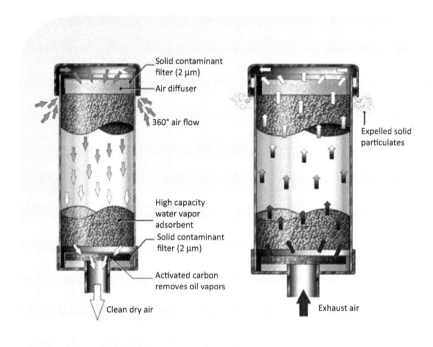

FIGURE 9.17
Desiccant breathers help control moisture ingress into oil sumps and reservoirs (Courtesy of US Lubricants, a division of US Venture, Inc.).

easiest way to control moisture is to proactively prevent it from getting in through the use of proper shaft seals (e.g., mechanical seals vs. simple lip seals) as well as the use of desiccant breathers (Figure 9.17). Desiccant breathers contain a silica gel desiccating medium that helps to remove moisture from air that enters the reservoir or oil sump through volumetric oil exchange or thermal cycling of equipment. Water can be detected using oil analysis or with a simple visual inspection of the sight glass or bottom sediment and water bowl. A desiccant breather can also be used as a CBM tool. A color change from blue to pink within the silica gel from the bottom up indicates external moisture ingress, while a change from the top down indicates moisture within the oil sump or reservoir.

Lubricant Contaminated with Particles

Most lubrication experts agree that as many as 60%–70% of lubrication-related failures can be tied directly to particle contamination. Just like water, the best way to control particles is through proactively restricting their ingress through proper shaft seals and breathers, as well as proper new oil storage, handling, and transfer. In addition, any circulating oil system should include properly selected filters, while noncirculating systems such as pumps and gearboxes should be periodically filtered using off-line kidney loop filtration (either portable or permanent). Filters should be specified with the correct micron beta rating to achieve the desired (required) target cleanliness rating, which should be determined based on equipment criticality and sensitivity to contamination-induced failure. Any proactive lubrication program should use oil analysis as a condition monitoring technique to determine whether the amount of particulate in the oil is below targeted levels for contamination.

Lubricant Degraded

When an oil or grease has been left in service for too long, both the base oil and additive performance can be impacted. Moreover, some of the by-products of base oil and additive degradation can result in the formation of sludge, varnish, and acids, all of which can result in further lubrication-related problems. Perhaps the simplest way to avoid lubricant degradation is through periodic oil analysis. Oil analysis can be applied to both oil and grease, though obtaining a representative grease sample can be challenging. Basic tests such as kinematic viscosity, acid number, oxidation, and nitration should be run to gauge the health of the lubricant as well as more sophisticated tests that measure base oil and additive health where appropriate. Used properly, oil analysis can be used to drive condition-based, as opposed to time-based, oil changes.

Lubricant Too Hot/Cold

Lubricant selection requires that the correct base oil viscosity be chosen for the operating temperature of the machine. But when the temperature of the lubricant becomes too hot or too cold, serious problems can occur. Too low a temperature during either start-up or unusual operating conditions can result in an oil or grease that's too viscous to flow to the load zone, resulting in lubricant starvation. Too high a temperature, and the base oil viscosity may be too low, resulting in 2-body abrasion, adhesion, leakage, and increased sensitivity to particle contamination. Oil temperatures should be monitored as part of the CBM program to ensure that the base oil viscosity is appropriate at all operating conditions, while the use of high–viscosity index (VI) lubricants such as premium mineral, multi-grade, and synthetic oils should be considered where operating temperatures are very high or low or vary widely throughout normal operation.

Additives Depleted

Oil additives help to control base oil oxidation, prevent rust and corrosion from occurring, prevent foaming and aeration, and prevent wear. As such, additive health is critical to proper lubricant function. Additive health can be measured using specific oil analysis tests that compare additive content from in-service oil samples with a representative new oil sample and should be included as part of routine oil analysis of large sump systems. Oil changes and regreasing schedules should be set to ensure that fresh oil or grease is applied in advance of additive breakdown or depletion. In rare cases, catastrophic additive depletion can occur through sudden water ingress or overaggressive filtration, both of which should be monitored carefully.

Lubricant Contains Foam/Air Entrainment

The presence of air in the form of foam and air bubbles in a lubricating oil can cause significant problems. Foam can result in leakage from sumps as the volume increases, as it serves as an insulating medium, causing oil temperatures within the oil sump to increase. Whenever foam or tiny entrained air bubbles are pulled into the suction side of a pump or into a bearing, the sudden pressure change can cause cavitation and micro-dieseling effects as compressive heating causes localized temperatures to increase. The presence of foaming and aeration can be detected using simple visual inspection, supplemented by more detailed laboratory-based oil analysis tests that measure foaming tendency and stability and air release characteristics of the oil.

Oil Analysis as a Predictive Tool of Other Problems

In addition to measuring lubricant health and condition, oil analysis is an excellent complement to other CBM tools such as vibration analysis and thermography. This is perhaps no more clearly illustrated than by a 2002 study from the Palo Verde Nuclear Plant in Arizona, which

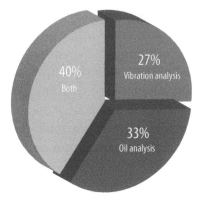

FIGURE 9.18
Effectiveness of oil analysis and vibration analysis in identifying common bearing failure modes.

concluded, based on 750 observed bearing problems, that as many as 40% of problems show up exclusively in oil analysis (Figure 9.18).[1] The point of this study was not to suggest that oil analysis is any better or worse than vibration analysis, but rather, to highlight the complementary nature of the two technologies and to illustrate the need for an integrated approach to predictive maintenance, deploying the most appropriate technology that addresses the commonly known or anticipated failures modes of any asset based on the results of the RCM-R® analysis.

At its core, oil analysis does more than just measure the condition of the oil; it proactively helps to measure the level of particles, moisture, or other contaminants that are present; it helps to predict the early onset of machine wear, including failure modes that are not directly tied to lubrication; and it helps to ensure that the base oil and additives contained within the oil are still healthy and that the lubricant is good for continued use.

To maximize the benefits of oil analysis, sampling frequencies, sampling locations, and test slate selections should all be tied back to the most common failure modes identified through the failure mode, effects and criticality analysis (FMECA) phase of an RCM-R® analysis.

Sampling Frequency

The P to F interval for most lubrication-related failure modes is anywhere from 14 to 180 days (or shorter in the case of lubricant starvation or wrong lubricant addition), meaning that oil samples should be taken fairly frequently, supplemented by basic visual inspection for oil level, color, and clarity. As a general rule of thumb, higher-speed, more critical applications should be sampled no less often than every month, while slower-turning, less critical, or intermittent-use assets can be sampled every 3 months, and visual checks should be done at least weekly and preferably daily. In most cases, sampling at longer intervals than quarterly is highly unlikely to yield positive results and will provide few, if any, data points for trend analysis.

Sampling Location

Where the sample is taken from has perhaps the most profound impact on the efficacy of oil analysis. For example, consider trying to use oil analysis to diagnose pump failure in a simple hydraulic power pack containing an oil reservoir and pump, a full flow oil filter on the supply line to the valve block and actuators, and a return line filter upstream of the oil return to the tank. Sampling from the reservoir, which is the most common location, is highly unlikely to diagnose pump failure, since wear debris would need to travel through the supply and return lines' filters and accumulate in sufficient concentration in the tank to permit an early indication of impending pump wear. Instead, the best way to address this specific failure mode would be to extract a sample immediately after the pump, before the full flow filter. In setting up an oil analysis program, careful consideration should be paid to the failure modes identified to ensure that the sample is being taken from the correct location. In some cases, this might require that multiple samples be taken from the same system: a process analogous to taking multiple readings such as axial and radial readings using vibration analysis.

Oil Analysis Test Slate

The series of tests performed on each sample should also be selected based on the known or anticipated failure mode. For example, a common failure mode in slow-turning gears is adhesive wear caused by heavy loads and/ or loss of boundary lubrication protection. When adhesive wear occurs, the initial wear particles formed are in the 10–20 micron size range. By comparison, elemental analysis, which is used in oil analysis to determine the concentration in parts per million (ppm) of specific wear metals such as iron, copper, lead, and tin, is completely blind to particles greater than 3–5 microns. As a result, simply applying a basic series of oil analysis tests to samples from a gearbox without including tests that can find larger particles will in all likelihood be unable to identify early-onset adhesive wear before it becomes too late.

ULTRASOUND

Allan Rienstra

Ultrasound testing is an important component of condition monitoring (CM), contributing value to RCM in many ways. Here, we will look at a short history of the study of ultrasound and its diverse applications, the basic principles of sound, and how to apply it to industrial applications.

Ultrasound is one of three words used to categorize sound waves. The other two are *sound* and *infrasound*, as shown in Figure 9.19.

- *Infrasound* encapsulates all sound waves below the limit of human hearing: that is, sound waves having frequencies below 20 Hz.
- *Sound* references sound waves perceptible to the human ear—that is, frequencies between 20 and 20,000 Hz.
- *Ultrasound* categorizes all sound waves above the human limit—that is, frequencies above 20,000 Hz.

The frequency of a sound wave is a measure of the number of times the sound wave repeats itself in one second (cycles per second). The

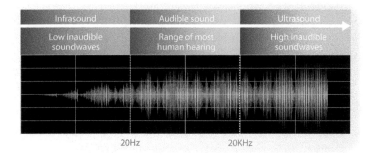

FIGURE 9.19
Sound waves: Frequency bands.

International System of Units (SI) named this measure after Heinrich Rudolf Hertz and assigned it the SI symbol Hz.

People began innovating with ultrasound as early as the eighteenth century, when it was discovered that bats used high-frequency sound waves for navigation. In 1881, Pierre Curie reported his discovery of the piezo-electric effect. His work remains relevant today, as the large majority of ultrasound technology uses piezoelectric crystals for both transmitting and receiving ultrasound waves.

Today, ultrasound technology has many well-known and practical uses. In the field of medicine, it serves as an imaging tool to enable physicians to noninvasively explore the human body. Pediatricians can monitor unborn babies inside the mother's womb. Oncologists use ultrasound imaging to search for elusive cancer cells. It contributes to human health with life-saving implications for the patient. Ultrasound is also helpful in sports medicine, where it acts on soft tissue injuries to speed healing and recovery.

There are many uses for ultrasound technology in industry as well. Nondestructive testing (NDT) uses super-high ultrasound frequencies (2–5 MHz) to perform imaging on physical structures. NDT reveals cracks in structural steel and flaws in welded joints, and measures the remaining thickness of steel pipes, metal plates, and ship hulls. Ultrasound is even useful as a parts and jewelry cleaning technology.

An important role for industrial ultrasound, and the focus of this chapter, is tied to asset reliability for manufacturing. Here, ultrasound serves many purposes, and each is tied to maximizing the life cycle of machine systems. Ultrasound provides a better understanding of the health of an organization's assets. It is used to identify and reduce energy waste. And when data is used to its fullest, organizations realize improvements in their output and product quality.

For the remainder of this chapter, all reference to ultrasound, unless otherwise stated, is focused on its use as a predictive maintenance, condition monitoring, troubleshooting, and energy conservation technology.

Sound Principles

When we discuss the principles of sound, the word *sound* is used, but the principles apply to infrasound and ultrasound inclusively.

Sound is ubiquitous. All things, human or machine, produce sound, and most beings can hear it. Those with hearing impairments may not hear sound but can still perceive its vibrations. Although sound is all around us, and therefore quite common, that does not mean it is a simple concept to understand. It is quite the opposite, in fact.

An important word used to describe sound is *frequency*, and that word is sometimes substituted with the word *pitch*. "The singer's rendition of *Silent Night* was *pitch perfect*" means that each note sung was perfectly in tune with the intended frequency of the sheet music.

There are often comparisons drawn between ultrasound testing and vibration analysis, which creates additional confusion. In ultrasound, the term *frequency* refers to the repetition of a sound wave, whereas in vibration, it refers to the repetition of a specific event, such as an impact from a defect or movement from machine imbalance.

At least in musical terms, that confusion is avoided by the use of the words *pitch* and *beat*, which adequately differentiate the frequency of the music from the rhythm of the music. Understanding the basic principles of sound serves to clear up several misconceptions.

The Basics

Sound is a mechanical wave that requires a medium through which to travel. It is produced by a vibration, which creates waves of pressure that transmit longitudinally through a medium. The medium supports the transmission of sound pressure waves and may be a solid, a liquid, or a gas, or any combination of these. Longitudinal waves are so called because they move through the medium in the same direction as the sound wave. As they move through the molecular structure of the medium, both high and low areas of pressure are created. These fluctuations in pressure are what produce sound that can be detected and measured.

Sound pressure waves propagate through a medium by molecular impact. To help visualize this concept of sound movement, imagine balls on a billiard table. When one ball collides with another, there is a transfer of energy passed on to the next ball, and so on. Other surrounding media, such as the cloth of the pool table, the side bumpers, and the surrounding air, all act on this energy to attenuate it in such a way that the balls eventually stop rolling.

A sinusoidal signal is a sound wave in which there is only one frequency present. During one complete cycle, there is a compression and a rarefaction of the signal, as illustrated in Figure 9.20.

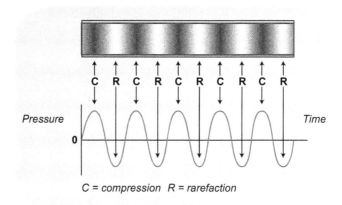

FIGURE 9.20
Sinusoidal sound wave showing compression and rarefaction components.

Ultrasound, as defined previously, is any sound pressure wave with a repetition frequency higher than 20,000 Hz. The frequencies most suitable for asset health monitoring and reliability of machine systems lie between 30,000 and 40,000 Hz. To quantify or measure sound, it is necessary to capture two variables: the frequency content present in the sound and the amplitude of the sound.

Frequency (F)

Frequency is a word used to describe how often an event is repeated. The frequency of a sound wave, in hertz, is the number of times the wave repeats itself in one second. A complete wavelength consists of one full compression and one full rarefaction event. Consider the use of the word frequency to describe these situations:

- The frequency with which one eats is three times per day.
- The frequency at which a data collector surveys plant assets is once per month.
- The frequency of a bicycle tire is 120 rotations per minute.

Each use of the word describes several events with occurrence intervals that can be measured. In the third example, the rotational frequency of a bicycle tire, the tire does two full rotations per second (120 rotations in 60 s). Therefore, the frequency of the bicycle wheel is 120/60 = 2 Hz.

Period (T)

Period (T) defines the length of time required for something to happen one time. It is inversely related to frequency (F). Frequency tells us how often something happens each second, whereas period tells us how long it takes to happen.

Period is the reciprocal of frequency and is calculated as $T = 1/F$.

To carry on with the bicycle wheel example, if 120 revolutions per minute is a frequency of 2 Hz, then the period of the bicycle wheel is 0.5 s.

It is important to understand this basic relationship between frequency and period. Ultrasound wavelengths occur with higher frequency than sound and infrasound wavelengths, and therefore have a shorter period, as illustrated in Figure 9.21.

Measuring Sound

Like most things, sound is measured on a scale. The scale for measuring sound is the decibel (dB). It is important to understand that the decibel scale is a ratio, not an absolute value. This means that an engineering unit is used as a reference against the decibel. The engineering unit chosen as the reference is dependent on the technology being used. For measurement of acoustics in a room—audible sound—it is common to see the SI unit dB (SPL) employed (sound pressure level).

Since decibels are a relative ratio, everything is measured against 0 dB (the threshold of hearing). dB (SPL) is often written lazily as just "dB,"

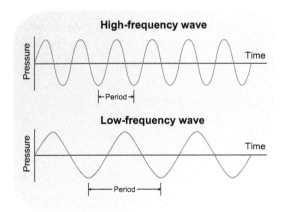

FIGURE 9.21
Relationship of period and frequency.

which misleads people to believe that dB by itself is an SI unit. It is important to clear that confusion, just as it is important to clarify that the decibel is a logarithmic scale, not a linear one.

In the world of ultrasound, the engineering unit used is the dBµV (decibel/microvolt). All measured amplitudes are compared with a reference value of 1 µV. Some ultrasound manufacturers publish this reference value openly, while others keep it as a guarded secret.

$$0 \, dB = 1 \, \mu V$$

SDT* instruments measure ultrasound using the formula

$$dBV = 20 \log_{10} \left(V_1 / V_0 \right)$$

where:
    V_1 is the measured voltage for the sensor in microvolts
    V_0 is the reference voltage of 1 µV

Let's use this example to make the point clear:

Assume that placing an RS1 needle contact sensor on a bearing housing produces an ultrasound signal with a measured voltage of 10 µV (V_1). Remembering that the reference value of the ultrasound data collector is $0 \, dB = 1 \, \mu V$ (V_0), we can say that the measured value from the bearing is greater than the reference value of the detector by a factor of 10. So,

$$dB\mu V = 20 \log_{10} \left(V_1 / V_0 \right)$$

$$dB\mu V = 20 \log_{10} \left(10 \, \mu V / 1 \, \mu V \right)$$

$$dB\mu V = 20 \times 1$$

$$= 20 \, dB\mu V$$

As this bearing degrades and enters failure, it will vibrate more, producing higher input voltages to the ultrasound data collector. Let's say that future measurements taken with the RS1 needle contact sensor produce input voltages (V_1) that are 10×, 100×, and 1000× greater than the reference value (V_0). What pattern will develop?

---

* SDT is a manufacturer of ultrasonic measuring equipment.

An ultrasound signal that measures 100 µV, 100 times the 1 µV reference value = 40 dBµV

$$dB\mu V = 20 \log_{10}(100 \ \mu V / 1 \ \mu V)$$

$$dB\mu V = 20 \times 2$$

$$= 40 \ dB\mu V$$

An ultrasound signal measuring 1000 µV, 1000 times the 1 µV reference = 60 dBµV

$$dB\mu V = 20 \log_{10}(1000 \ \mu V / 1 \ \mu V)$$

$$dB\mu V = 20 \times 3$$

$$= 60 \ dB\mu V$$

0 dBµV is the threshold of "hearing" for any of SDT's ultrasound instruments. Knowing this base is imperative to establishing trends and interpreting condition indicators. Without that knowledge, an ultrasound inspector is essentially blind (deaf?).

Maybe the most important rule to differentiate linear from logarithmic is that decibels should never be multiplied or divided. They should only be added or subtracted. For example, 36 dBµV is not 2× louder than 18 dBµV. Instead, subtract the difference (36 − 18 = 18 dBµV). 36 dBµV is therefore louder than 18 dBµV by a factor of 7.9.

Table 9.2 lists some common ratios and their relationship to the dBµV. The first one demonstrates that any increase of 6 dBµV over a previous value represents a signal that is two times louder (double).

TABLE 9.2

Signal Ratio vs. dB Rating

Ratio (x)	dBµV
2	6
4	12
10	20
100	40
400	52
1000	60

To drive the point home,

- 12 dBμV is two times louder than 6 dBμV.
- 56 dBμV is two times louder than 50 dBμV.

Velocity of Sound and Acoustic Impedance

There is a relationship between the speed at which sound travels through its medium (V), the number of times it repeats itself (F), and its wavelength (λ). This relationship is defined by the equation

$$v = \lambda f$$

The velocity of sound through a material is dependent on many variables, such as the elasticity of the medium and its density.

Some examples:

- Sound travels roughly four times faster through water than it does through air.
- Sound travels roughly 15 times faster through steel than it does through air.
- Sound travels roughly three times faster through helium than it does through air.

Understanding the velocity of sound through a medium helps relate to the acoustic impedances of different materials. Acoustic impedance describes how a material resists the propagation of sound through it. Acoustic impedance (z) is the product of density (p) and velocity (v):

$$z = pv$$

In the world of ultrasound inspection, respecting the behavioral effects of acoustic impedance leads to more accurate data. Inspectors must understand what happens to an ultrasound signal when it passes from the boundary of one material into another. Every change in material from the sound's source to its destination is impacted by the acoustic impedance of the medium, including loss of signal energy.

Reliable CM data is necessary for trending, alarming, interpretation, and decision-making. Data is obtained through direct contact between the piezoelectric sensor and the asset's surface. To ensure integrity of the data, it must be captured through the lowest number of boundary changes.

An example where boundary behavior control is mandatory is monitoring bearing condition. Data collected on the bearing housing may pass through as many as three or even four different media. Dirt and paint both count as a boundary and must be avoided. Taking the bearing measurement on the grease pipe is a more direct path, but one must still be wary. Is the grease pipe made of steel (acoustic impedance 4.516) or is it aluminum (acoustic impedance 1.71)? What if it was originally steel, and then later replaced with aluminum, but no one bothered to document it? Up to 20% of the ultrasound signal could be compromised. This change alone would reduce the transmitted signal by 1.6 dBμV and completely destroy any historical trend data.

Some good rules to follow for data with high integrity:

- Limit the number of boundary transitions from source to sensor.
- Collect data at the same point all the time.
- Collect data on clean, unpainted surfaces where possible.
- Use permanently mounted resonant sensors.
- Alternatively, use magnetic mounted resonant sensors.

Sound Propagation through Air

Air is a transport medium for ultrasound with its own share of quirks. Understanding how ultrasound travels through air is a mandatory element of every ultrasound inspector's apprenticeship. Consider once again the journey from source to sensor, but this time through air instead of solids. The longer the distance traveled, the more signal energy is lost. This poses a problem for quantifying airborne ultrasound. It is imperative that the distance between the sensor and the noise source be known and constant.

Sound attenuation as a function of distance is described by the inverse distance law. This law states that the measured pressure (p) changes inversely with distance (r). The formula is written as

$$p \propto \frac{1}{r}$$

As a result of this, when the distance between source and sensor is doubled, the amplitude of the signal is halved. This represents a reduction in measurement amplitude of 6 dBμV. Halving the distance will produce the opposite effect. The measurement will increase by 6 dBμV or a factor of 2.

Take as a comparison what happens to the amplitude of the same sound source when measured at various distances. If a sound source has measured amplitude of 60 dBμV at a distance of 10 cm, that same sound source would measure

- 54 dBμV from 20 cm away
- 48 dBμV from 40 cm away
- 42 dBμV from 80 cm away

If the detection of airborne ultrasound signals requires measurement of amplitude, then it is important to know and record the distance for consistent and comparative measurements.

How Ultrasound Detectors Work

It would be an oversimplification to say that ultrasound detectors have a singular purpose of converting inaudible ultrasound into audible sound. However, that is precisely what they do. The process of shifting a frequency is called *heterodyning*, illustrated in Figure 9.22. When a high-frequency soundwave is heterodyned to an audible sound wave, the quality and characteristics of the original signal are maintained. In effect, an ultrasound detector equips humans with the ability to hear ultrasound. How well the quality and characteristics of the signal are maintained during heterodyning speaks to the quality of the ultrasound detector.

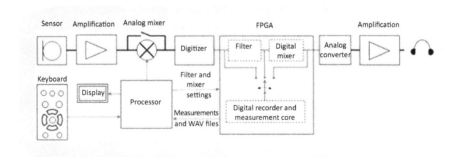

FIGURE 9.22
Typical ultrasound detector's processing layout.

Following the simplified block diagram of Figure 9.21 from left to right, top to bottom, an ultrasound signal goes through a series of steps before it is useful for maintenance and reliability:

1. The sensor detects the sound pressure wave and converts it to a low-voltage signal.
2. That signal goes through a controlled amplification process.
3. The analog mixer converts the high frequency to a low frequency (heterodyne step).
4. The signal is digitized.
5. Bandwidth filters are applied.
6. The signal is split at this point:
   a. The analog portion is sent to the audio output (our ears).
   b. The digital signal is used to accurately measure the signal.
7. Both static and dynamic signals are created for trending and time waveform analysis.
8. Results are sent to the liquid crystal display (LCD).
9. A keyboard is used to control the graphical user interface.

Figure 9.23 presents a more simplified representation. A 2 kHz bandwidth filter centered at 38.4 kHz heterodynes an ultrasound signal into its corresponding audible signal while preserving the characteristics and quality of the original.

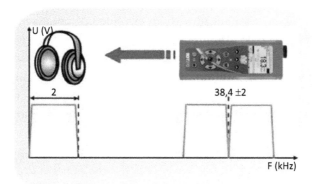

FIGURE 9.23
Heterodyning converts signal at 38.4 kHz into a 2 kHz audible sound.

How Ultrasound Is Collected

Manufacturers of ultrasound hardware commit significant resources to developing a variety of different sensors. It is generally accepted that there are two categories of sensors: one for detecting ultrasound through solid media and another for detection through air. Each is designed to fulfill a specific task.

Airborne Sensors

As their name suggests, airborne sensors detect ultrasound that propagates through air. There are several styles of airborne sensors, each capable of overcoming specific inspection challenges. Some are designed for close-range detection, others for far distances, while others permit inspection in tight access areas. Quality manufacturers offer sensors that pay respect to the following criteria:

- Form and function
- Accuracy, high signal to noise ratio, and repeatability
- Ruggedness and durability
- Ergonomics and safety

Sensors not meeting these criteria should not be used.

Close-range airborne sensors are designed for short and mid-range distances. These are mounted inside the housing of the ultrasound detector, thereby ergonomically placing the sensor and the readout in direct view of the inspector. The housing lends protection and robustness.

Internal sensors are enhanced with an extended distance sensor (EDS), as shown in Figure 9.24. This conical-shaped apparatus screws over the sensor and acts as an amplifier by capturing more sound pressure waves and funneling them onto the sensor. An EDS looks simple enough, but its complex design is anything but. Its throat, the mouth opening, the length, and the profile of the EDS all determine the sensitivity and resonance in a specific and narrow range of frequencies.

SDT, a manufacturer of ultrasound systems, cites an amplification gain factor of 20×, or 26 dBµV, with its EDS. The benefits are increased distance of detection as well as the ability to detect signals that are 20× quieter. Reception is much more directional with the EDS fitted over the internal sensor than it is without.

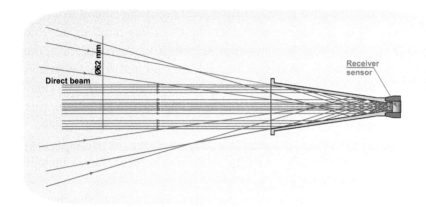

FIGURE 9.24

Extended distance horn (EDS) enhances receiver sensor for long-range use and detection of "quieter" signals.

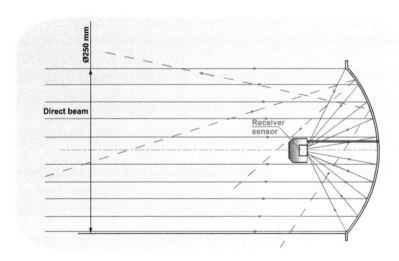

FIGURE 9.25

Parabolic sensor for directional accuracy.

Directionality is a desirable characteristic that helps inspectors pin-point the exact source of the ultrasound wave over long distances. EDSs should not be considered a replacement for parabolic sensors, shown in Figure 9.25. While they both extend the distance of detection consider-ably, the EDS amplifies signals from the entire field of detection.

The shape of the parabolic sensor is engineered, like the EDS, so that sound waves of a specific frequency are guided onto the center axis of the sensor. Audible sound waves are too large to reflect off their surface. A parabolic dish is designed to reflect incoming plane wave signals to a focal point where one or multiple sensors are mounted. Ultrasound signals that approach the dish at angles that are not perpendicular to the plane of the parabola will not reach the sensor. Instead, they are redirected at angles that reflect away from the sensor's focal point (see Figure 9.25). This characteristic makes a parabolic sensor the most appropriate for pinpointing the source of an ultrasound signal over very long distances (100–200′+).

In applications where pinpoint accuracy is important, the parabolic dish provides much greater directionality, and most manufacturers include laser sighting.

Some inspections require access into hard-to-reach areas, including behind piping, overhead, underfoot, and through protective guarding (flexible couplers, for example). Long, tedious inspections in electrical switch gear rooms can become tiresome for the inspector's hand and arm. A flexible sensor, designed ergonomically, gives the option to holster the ultrasound detector so only the comfortable, foam-gripped handle need to be held up.

Flexible sensors are an extension of the internal sensor. They typically add around 20–26 in of length and can be bent, twisted, and shaped to fit behind pipes. The use of flexible sensors provides ergonomics but more importantly, an element of safety.

Contact Sensors

The other category of detection is ultrasound that is structure borne through solid media. This is realized with a contact sensor, as shown in Figure 9.26. There are two styles of contact sensors: needle or stinger

FIGURE 9.26
Contact sensor.

type (shown in the figure with its needle at the bottom) and permanent mounted or magnetic mount.

Structure-borne ultrasound waves are produced inside a body such as a bearing, a steam trap, a gearbox, a valve, or another mechanical system. Contact sensors act like an antenna for structure-borne ultrasound signals. Their purpose is to form a transfer medium between the ultrasound source and the sensor. For repeatability and efficient transmittal of the ultrasound signal, resonant contact sensors must have fixed-length needles. Sensor manufacturers have studied the optimum lengths for transmitting these high-frequency, low-energy signals.

Some ultrasound guns are designed only for troubleshooting. For this style of instrument, where measurement accuracy and data integrity are a lower priority, it is possible that some design rules are not observed. For ultrasound detectors that are designed to detect and measure ultrasound signals, more emphasis is placed on contact sensor design.

The frequency response of a sensor is always closely matched to the instrument. Sensors should be interchangeable. This means that regardless of which ultrasound instrument it is plugged into, the data is verifiably trusted. Likewise, when a sensor breaks and is replaced, the new sensor must closely match the performance of its predecessor.

The question of when to use the frequency filter tuning is often asked. Most mechanical problems that ultrasound programs solve generate peak frequencies in the 36–40 kHz range. Resonant contact sensors use 40 kHz piezo crystals. It is possible to use bandwidth filter tuning to adjust the ultrasound instrument to listen to other frequencies. However, this should be done with caution, as the sensor and instrument calibration only certifies accurate measurements at its optimum frequency. Frequency tuning is best reserved for troubleshooting, not trending.

Why Ultrasound Is an Effective Technology

The inherent characteristics of ultrasound waves are what make ultrasound such an effective technology for detecting machine system defects in noisy, industrial environments.

Ultrasound signals have short, low-energy wavelengths. As such, they do not propagate well through their medium. This is actually an advantage for inspectors, because the defect's signal is loud at its source but attenuates after a short distance. Trying to detect problems using audible signals is confusing, since competing sounds from nearby machines make it impossible to distinguish where the bad actor resides.

Another characteristic of ultrasound is directionality. A low-energy soundwave is constrained within its medium. Remember that ultrasound waves are longitudinal; their propagation path is strongest in the direction they are cast. Waves move through their medium by molecular impacting, but the signal's energy is too weak to fan out. In fact, a 40 kHz ultrasound signal suffers significant energy loss beyond a 60° path.

The advantages offered by directionality are many. Consider that a simple machine system such as an electric motor driving a pump will have as few as four bearings and a flexible coupling. Directional ultrasound allows data to be received from all five measurement points without fear of competition.

The same advantage exists for leak detection. A poorly maintained compressed air system may see up to 40% of total demand lost to leaks. Several leaks may exist in close proximity, but thanks to the directional nature of ultrasound, it is possible to find and pinpoint individual leaks in these conditions.

Ultrasound allows maintenance and reliability to hear above the noise of the factory floor. By screening out low-frequency sounds, inspectors hear sounds consistent with machine system defects. These defects produce ultrasound from three phenomena:

- Friction (F)
- Impacts (I)
- Turbulence (T)

When faced with the decision whether ultrasound is a suitable technology to find certain failure modes, ask yourself "Is it FIT?" If the defect produces friction, it's a fit. If it produces impacting, it's a fit. And if it generates turbulence, it is also a perfect FIT for ultrasound testing.

Applications

As a CM technology, ultrasound has hundreds of useful applications in almost any manufacturing sector. It is most often used for bearing CM, bearing regreasing, analysis of low-speed rotating assets, steam trap testing, valve bypass, electrical discharge detection such as corona, arcing, and tracking, compressed air leak management, and discovering leaks in shell and tube heat exchangers. It is a powerful CM technology that maintenance professionals find extremely useful.

Determining where, and where not, to employ ultrasound technology is as simple as following SDT's "FITness Test." Machine system defects produce ultrasound from FRICTION, IMPACTING, and TURBULENCE. Simply match the defect to the phenomena to determine the suitability of ultrasound as a solution.

Compressed Air Leak Management

Compressed air is one of the top three most expensive utilities used in manufacturing. Leaks are expensive, and often ignored. While they can be heard with the naked ear, they are difficult to pinpoint because of background noise. An ultrasonic detector hears leak TURBULENCE above the ambient noise of the factory floor.

The high-frequency component of a leak is directional, making it fast and easy to locate its source. A compressed air survey with an ultrasonic detector once per quarter reveals savings potential in the millions and benefits facilities managers looking to improve efficiency and reduce costs.

Condition Monitoring

CM of rotating and nonrotating equipment continues to evolve alongside the new generation of detectors. Assets produce FRICTION and IMPACTING with high-frequency ultrasonic signatures that have peaks at 35–40 kHz. Friction is masked by ambient plant noise and low-frequency vibrations, but is clearly heard and measured in the ultrasound range. Changes in these signatures serve as early indicators of failure and provide comparative, complementary information for vibration analysis.

Slow-Speed Bearings

Slow-speed bearing monitoring presents a challenge for seasoned vibration analysts. Common defects include pitting, impacting, and rubbing. All produce friction and impacting, best viewed with dynamic data using time wave analysis tools.

Acoustic Condition–Based Lubrication

Machines depend on proper lubrication to reduce frictional forces, which otherwise shorten the asset life cycle. Optimized lubrication means applying new grease only when it is needed. It also means using just the right amount of grease to return friction levels to an acceptable level.

Ultrasound provides data that encourages maintenance teams to shift from calendar-based lubrication tasks to on-condition scheduling. It warns when friction levels elevate above an acceptable baseline, and it guides as just the right amount of grease is applied and levels return to normal.

Electrical Applications

The versatility of ultrasonic inspections extends to the electrical maintenance department, where routine scans of switchgear, substations, and high-kV transmission and distribution lines are commonplace. There is mounting concern about safety, and specifically, the danger of arc flash. Prior to opening high- and medium-voltage electrical panels, inspectors use ultrasound detection to listen to the levels of ultrasound inside the cabinet.

Steam System Inspections

A steam trap is an automatic valve that opens for condensate and non-condensable gases and closes for steam. It is designed to trap and remove water, air, and CO_2, which hinder the efficient transfer of steam, corrode system components, and cause damaging water hammer.

Ultrasonic surveys of the entire steam system will reveal system leaks, blockages, stuck valves, and failed traps. Increasing steam efficiency translates to huge dollar savings and increased product quality.

Certain types of traps can benefit from dynamic signal analysis. When monitoring continuous traps, it can be difficult to discern between live steam from a failed trap and flash steam, which is produced when the reduced pressure of the condensate line causes condensate to regenerate back to steam. Viewing the time signal of suspect traps can help distinguish between flash and live steam.

Pump Cavitation

Cavitation is the result of a pump being asked to do something beyond its specification—draw from too low a suction pressure. Small cavities of vapor develop behind the vanes. These pockets impact destructively on the pump's internal components. Cavitation damage may range from minor pitting to catastrophic failure.

During normal data collection, inspectors use ultrasonic detectors to isolate random cavitation, which can be masked by low-frequency modulations. Early detection with ultrasound and preventive solution implementations will prevent long-term damage and unnecessary downtime incurred from cavitation issues.

Reciprocating Compressors and Valves

Reciprocating valves give breath to compressors. Worn or dirty valves can't seat properly. Over time, springs weaken, limiting the force necessary to snap open and closed, and causing leakage. Valve condition is monitored with ultrasound inspection and spectral analysis software. Spectra graphs visualize the compressor valve as it opens and closes, and intakes and exhausts.

Visualizing the recorded sound file of a compressor valve in the time domain tells us a lot about the condition of the valves and their components. There are three distinct events (open, intake or exhaust, and closed), all occurring at split-second timing, far too fast for human ears to process. By viewing the wave file in real time, we can stretch it out to visualize each individual event.

Reciprocating compressors produce ultrasound through all three phenomena (FIT):

- Valve opens: Friction and impacting
- Intake or exhaust: Turbulent flow
- Valve closes: Friction and impacting
- Valve leaks: Turbulent flow

Heat Exchanger and Condenser Leaks

Tube condensers and heat exchangers cool steam, which condenses back to purified water and is returned to a boiler, where it's superheated back to steam. Leaks in exchanger tubes allow feed water to leak out or contaminants to enter, allowing corrosion and reduced operating life. Keeping the water pure is the key to efficiency.

The general method of inspection involves scanning with the instrument a couple of feet from the tube sheet, noting any noisy areas. Then, the inspector switches to an extended flexible sensor to scan tube to tube. If the sound signal on the digital dBμV meter or sound in the headset does not change from tube to tube, a leak is unlikely. If a significant signal change occurs, then a leak is suspected. If the leak is within the tube, the difference will be heard at the tube opening. If the noise level is heard on the tube sheet, block the area to eliminate reflected noise. Use a precision tip with an opening of one eighth of an inch on the flexible extended sensor, and hold it almost on the tube sheet surface to pinpoint leak locations.

Ultrasound for Reliability

The field of ultrasound is an interesting one, to be sure, with countless applications including navigation, medicine, materials testing, and industrial CM. Although ultrasound testing has been around for 40 years, applications specific to industrial maintenance, and specifically CM, continue to expand.

Ultrasound for reliability is a technology that can be both simple and complex, depending on its deployment. A compressed air leak management program or a steam systems survey can be fast and easy to implement and net tremendous return on investment. Inspecting electrical systems for potential faults increases safety and awareness with lifesaving potential. Monitoring rotating and nonrotating equipment for degradation and wear warns maintenance when a machine's life cycle is about to be cut short. Using ultrasound data to decide when motor bearings need relubrication, and then further guiding lubricators to prevent overgreasing, reduces downtime, saves on wasted grease and oil, and frees up manpower for more meaningful work.

Ultrasound programs contribute to reliability in a significant way. For organizations in pursuit of a complete RCM philosophy, a world-class ultrasound program should be included.

NONDESTRUCTIVE TESTING

Jeff Smith

A key objective of our reliability programs is to cost-effectively manage the overall integrity of assets. RCM-R® reveals many ways in which asset integrity can become compromised, and CBM targets the onset of failure for many assets. Some assets, however, are not conducive to the application of nonintrusive CBM methods such as vibration analysis, thermography, oil analysis, and ultrasonic inspections. In these other cases, we can use nondestructive testing (NDT) to help us manage the failure mechanisms of those other assets.

NDT is helpful for failure mechanisms such as corrosion, erosion, rupture from bursting, and external forces. The integrity of assets such as piping, vessels, tanks, boilers, and many others is predicted and protected through the use of NDT.

Early industrial history reveals a poor track record of managing stationary assets. They would burst, corrode away and leak, collapse, and so on. Health and safety were often put at risk, and in some cases, there were fatalities and injuries. There are many historical examples of piping, boiler, and tank failures that have contributed to loss of life and environmental damage. Governments stepped in to protect workers and the general public, resulting in the requirement for standards and governance to ensure public safety; however, even today we experience avoidable accidents.

Some recent examples are

1. An explosion in San Juanico, Mexico, killed hundreds and injured thousands in 1984.
2. In Louisiana in 1988, a corroded pipe leaked, resulting in a hydrocarbon gas escape that ignited, killing 7 and injuring 42.
3. In Flixborough, England, in 1974, a feed pipe failed, resulting in an explosion that leveled the plant, killed 28, and injured another 36. There were also 53 civilians injured when the blast damaged 2000 nearby buildings.
4. In Bhopal, India, in 1984, an uncontrolled chemical reaction due to equipment failure released a toxic cloud that killed nearly 2,000 civilians and injured an estimated 20,000 more.
5. In Pasadena, Texas, in 1989, a massive explosion resulted in 23 fatalities and 315 injuries.

In addition to the application of other CBM techniques, NDT is a key program that directly impacts your health, safety, and profitability and the environment you work and live in.

Today, we have multiple industrial standards that were developed mandating NDT. There are multiple governing bodies that provide standards related to NDT, including the American Petroleum Institute (API), the International Organization for Standardization (ISO), and the Society of Automotive Engineers (SAE).

Many failures start as small cracks—undetectable to the unaided naked eye. There are a multitude of techniques used to conduct NDT for crack detection. These methods include visual testing, liquid penetrant inspections, and magnetic particle, hardness, and ultrasonic testing. Note that in most cases, these techniques cannot be used with the equipment in operation. NDT is often intrusive; nevertheless, it provides us with another suite of tools in our CBM arsenal.

Conventional NDT Methods

Visual testing is conducted by looking for surface anomalies, often with the use of optical enhancement tools such as microscopes, controlled cameras, borescopes, endoscopes, telescopes, high-speed cameras, and so on.

Magnetic particle detection is an enhanced crack detection method. Magnetic particle detection is conducted by dusting a ferrous metal surface with iron oxide particles and inducing a magnetic field in the material. Surface or near-surface flaws disturb the magnetic field and "leak" magnetic flux in the flawed areas. The iron particles are attracted to the flux leakages, producing a visual indication of the defect.

Liquid penetrant inspection is another crack detection method used for surface cracks. The object is coated with a solution that contains a visible or fluorescent dye. The excess solution is then removed and a developer applied; cracks in the surface will retain some of the solution, so they can be seen.

Radiography is used to inspect items internally and externally. Radiography uses penetrating radiation to examine materials' internal features. Radiation is directed through the part and onto a detector, much like an x-ray is used on the human body, only with much stronger x-rays. This process results in a shadow graph that displays material thickness, density changes, and voids.

Pulse-echo ultrasonic testing (UT) has become one of the predominant tools used for NDT. In pulse-echo ultrasound, an electromagnetic acoustic transducer sends and receives a pulsed sound wave. The sound wave will bounce back to a detector whenever it hits either the other surface or an anomaly within the material. This technology will detect thickness as well as corrosion and other included flaws such as cracks or holes in cast materials. Quite often, UT will be used on a set frequency and results trended to help project remaining useful life where erosion or corrosion is a major form of degradation. Trending also helps to detect any changes in rates of degradation.

Advanced NDT Methods

In addition to the conventional NDT methods, there are also a number of advanced tools and technologies, including UT crack detection and sizing, C-scan corrosion mapping and flaw detection, time of flight diffraction, phased array, guided wave UT, and digital radiography.

UT crack detection and sizing. The depth of a surface crack can be evaluated using the pulse-echo technique. Longitudinal waves are employed to investigate the effect of frequency on the sizing detection of surface cracks. Reasonable accuracies have been achieved, with measurement errors less than 7%.

C-scan corrosion mapping and flaw detection. Corrosion mapping is performed with an automatic or semiautomatic scanner. An inspection surface is scanned using various ultrasonic techniques including pulse echo, eddy current, and phased array. Corrosion mapping is widely used in the oil, gas, and nuclear industries for the inspection of piping, pressure vessels, tanks, boilers, and reactors.

Time of flight diffraction uses a pair of ultrasonic probes. The probes are set on opposite sides of a weld with a transmitter and a receiver. The transmitter emits an ultrasonic pulse that is picked up by the probe on the other side, the receiver. When a crack is detected, there is a diffraction of the ultrasonic wave from the tip(s) of the crack. Using the measured time of flight of the pulse, the depth of a crack tip(s) can be calculated automatically.

Phased array consists of many small ultrasonic transducers, each of which can be pulsed independently. The beam can be focused and steered electronically. The beam is swept through the object being examined, and the data from multiple beams are put together to make a visual image showing a slice through the object.

Guided wave UT. Guided wave testing uses very low ultrasonic frequencies, enabling the sound wave to travel along a pipe, providing 100% coverage of the pipe length. An array of low-frequency transducers is attached around the circumference of the pipe to generate a wave that propagates along the pipe in both directions from the transducer location. Evaluating the properties of the wave relies on heavy mathematical modeling, which is typically presented in graphical plots called *dispersion curves*.

Digital radiography is a form of x-ray imaging, where digital x-ray sensors are used instead of traditional film. This process can be carried out faster than conventional x-ray and uses less radiation to produce an image. Instead of x-ray film, digital radiography uses a digital image capture device.

Risk-based inspection (RBI) is a method for developing inspection programs that use NDT. It complements RCM-R® and can be used where NDT is indicated as a failure management policy, or it can be used on its own if you are confident that you've leveraged other CBM techniques as much as you can.

RBI uses assessed risk levels to develop a prioritized inspection plan. The RBI process evaluates the potential damage mechanisms of static equipment such as piping, pressure vessels, and heat exchangers. It considers the risks of both active and potential damage mechanisms against the business, environmental, health, and safety consequences of failure. The output of an RBI study provides optimized inspection frequencies and a definition of which type of NDT to apply. It can also be used to evaluate operational envelopes and loading.

The application of RBI safeguards asset integrity as well as improving the reliability and availability of the asset. RBI also tends to reduce the number of inspections as well as the requirement for shutdowns, providing a longer operational campaign without compromising reliability.

There are several standards that recommend and outline the requirements for RBI:

API 580: Risk-based inspection recommended practice
API 581: Risk-based inspection resource
API 571: Damage mechanisms affecting fixed equipment in the refining industry
ASME PCC-3: Inspection planning using risk-based methods

Like RCM-R® outputs that define a structured work program, RBI tasks must be implemented into the planned and scheduled work. During the execution of inspections, if data indicates the presence of a damage mechanism in a component, quick follow-ups must be performed to ensure that the best data (e.g., lowest thickness) was collected. If this is done in an almost real-time fashion, then follow-up work can take advantage of the already open access granted for the initial NDT inspection.

CONCLUSION

This chapter has presented a number of CM technologies of varying complexity. Knowledge of these is important to your RCM-R® analysis—it gives you a number of excellent options for CBM (C tasks). It is important to bear in mind that choosing any of these technologies brings with it a need to train and qualify your technicians in its proper use. As you can now see, each of these technologies must be applied correctly to produce the result you want—accurate forecasting of a potential failure condition. In untrained hands, these technologies can produce results that are easily misunderstood and then misapplied—potentially leading to corrective work where none is really needed (false alarms) or to a false sense of security if potential failures are undetected. Both of these can undermine confidence in your entire CM program. However, when used correctly and by suitably qualified and knowledgeable technicians, these technologies can produce remarkable results and findings that otherwise are likely to go unnoticed.

When properly applied, RCM-R® will produce a wide array of CM outputs. Combined with the five physical human senses, you now have a powerful toolbox full of excellent options—each with particular strengths for your particular failure mode causes in their operating context.

REFERENCE

1. Bryan Johnson Palo Verde Nuclear Power Generating Station, "Oil Analysis Success at a Power Station", *Machinery Lubrication Magazine*, July 1998, Noria Corporation, Tulsa, OK.

10

Selecting Strategies for Managing Failure Consequences

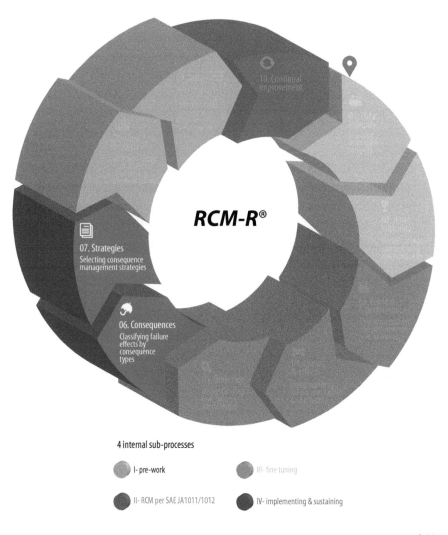

RCM-R®

07. Strategies
Selecting consequence
management strategies

06. Consequences
Classifying failure
effects by
consequence
types

10. Continual
improvement

4 internal sub-processes

I- pre-work

III- fine tuning

II- RCM per SAE JA1011/1012

IV- implementing & sustaining

CATEGORIES OF FAILURE CONSEQUENCES

Failure effects should describe and quantify the impact of every failure root cause on business objectives with regard to cost, safety, and environmental influences, help us identify whether the failure is hidden or evident, and possibly identify other impacts. They enable us to justify the consequence management policy for risk elimination in the current operating context of the asset. Other impacts may also be important. For instance, deterioration of the firm's reputation resulting in potential business loss may be a risk in some cases.

RCM-R® classifies failure effects into categories based on evidence of the failure's impact on safety, the environment, operational capability, and cost. P (Production), M (Maintenance), S (Safety and Environmental), and H (Hidden) are the four possible failure consequence categories the analysis team must use to classify failure effects. We use the information provided in the failure effects descriptive paragraph as written for each failure mode and cause, and the failure effects risk analysis explained in Chapter 7 to decide which of the four categories applies each time. Only one category is chosen for each failure mode cause—whichever reflects the most severe consequences.

> *P (Production)*: This category of failure is chosen when there is only economic impact associated with evident failures, and the costs of production losses exceed those related to repair activities. Failure events resulting in loss of raw material, production rework and quality defects, and increased energy and/or labor costs but not affecting safety or environmental goals fall into this category. Also, the cost of downtime associated with production deferment is considered here. This type of consequence is regarded as *operational* in SAE JA1011.
>
> *M (Maintenance)*: Evident failures having only economic impact, with repair or replacement costs surpassing those related to production losses, are categorized in the maintenance consequences category of effects. These include costs related to direct labor, materials, contractors, spare parts, and support equipment rental and logistics, among others. They are regarded as *nonoperational* in SAE JA1011.
>
> *S (Safety and Environmental)*: Evident failures having impact on safety and environmental goals are classified into this category. Therefore, failure events with the potential of causing physical or psychological damage to operators, maintainers, or the community as a whole are clearly highlighted for appropriate treatment in this failure

consequence class. The same rule applies to events having adverse impact on a firm's environmental-related goals.

H *(Hidden)*: Failure events that have already happened but haven't yet become evident to the operators under normal operating circumstances fall into this failure effects classification. These failures are almost always associated with machine protective devices that are normally inoperative, faulty redundant components, defective safety devices, and unavailable backup items. The worst-case scenario posed by a hidden failure is the occurrence of a multiple failure which will also have either safety, environmental, operational, or non-operational consequences. This is the case when the protective device is out of order when the protected function fails, as shown in Figure 8.3.

In Chapter 5, we explained that RCM-R® classified functions as primary, secondary, or hidden, and that failures are further classified as critical, non-critical, or hidden. *Hidden failures* occur when asset protective devices or redundant components fail to fulfill their function. They represent the inability of an asset to carry out its hidden functions effectively. Hidden failures are (by definition) not evident to the operator or maintainer of the machine during normal operation and may expose the organization to multiple failures affecting the integrity of people, the environment, machinery, or the product. On the other hand, *evident failures* are palpable to the asset's operation and maintenance personnel. The failure effects classification process clearly separates failures having only economic impact from those exposing the organization to failure to attain safety and environmental goals. Note that *hidden failures* may impact economic, safety, and environmental goals as described in their failure effects statements. The following step-by-step process (in the presented order) is found to be quite helpful in helping the analysis team to decide which failure consequence must be chosen for each failure effect:

1. Failures classified as hidden are considered to have type H consequence.
2. Effects of evident failures causing injuries or death to operating personnel or potentially triggering environmental incidents should be classified as having type S consequence.
3. Evident failures having only economic impact must be evaluated to determine whether production or maintenance cost (or budget) is more affected by the failure. Type P consequence is chosen whenever production losses or increased costs exceed repair or maintenance costs.

4. Type M consequence is then selected when the opposite is found for evident failures not having safety or environmental impact. That is, maintenance repair activities' costs exceed production-related costs due to the particular failure cause.

Alternatively, Figure 10.1 can be followed to classify all failure effects into their corresponding categories based on the evidence of failure impact on safety, the environment, operational capability, and cost. The next logical step in the RCM-R® analysis is the assignment of the failure effects consequence to each failure mode root cause, as shown in Figure 10.2.

FIGURE 10.1
Failure effects classification questionnaire.

FAILURE CONSEQUENCE MANAGEMENT POLICIES' NOMENCLATURE AND TYPICAL DECISION DIAGRAMS

The failure consequence choice paves the way for the selection of appropriate consequence management policies, often called *maintenance strategies*. The analysis team selects specific tasks with corresponding frequencies where applicable. In Chapter 8, we explained different consequence management policies we can consider for mitigating every critical failure cause. The important premise here is that the recommendations we make must be technically capable of dealing with the failure causes, and at the same time, they must be worth doing from a perspective of costs or risk mitigation.

RCM-R® classifies and assigns a specific code to failure consequence management policies by the type of tasks each one uses, as follows:

- C: Condition monitoring (considers predictive maintenance, nondestructive testing [NDT], and process parameters trending)
- T: Time-based restoration or replacement
- D: Detection
- O: Operator-performed tasks
- 2: Combination of two types of tasks among C, T, and D types.
- R: Redesign tasks (one-time changes as explained in Chapter 8)
- F: Failure (run to failure)

Condition monitoring, time-based and detection tasks are performed cyclically and are listed in preventive maintenance (PM) program job plans. However, they focus on dealing with different types of failures and failure patterns. Whereas C and T maintenance tasks try to avoid functional failures or reduce their consequences to a tolerable level, D type tasks are intended to find (hidden) failures that may have already occurred by the time the task is carried out. Type D tasks are used to reduce the risk of a multiple failure to a tolerable level. Type C tasks are used to find potential or incipient failures that may be tracked down through a condition parameter to avoid a possible functional loss of the asset, as explained in Chapters 8 and 9. T tasks are designed based on the useful life of the item and encompass the restoration or replacement of an item at or before the determined useful life.

Functions and functional failures with classification
(critical, noncritical, and hidden)

RCM-R®									Failure effects risk evaluation	Consequence
				Failure modes/ types & root causes/ type						
Function #	Function per formance level and operating context	#	Functional failure (loss of function)	Failures classifica-tion	#	Failure mode	Type #	Root causes	Type	Failure effects*	
										Total	
2	To supply a minimum of 30 gpm of caustic water	A	Unable to supply caustic water at all	C	1	Pump coupling wear	MAT C	Misalignment caused by wrong practices	MGT	The vibration alarm will alert the operator there is a problem with the pump. The mechanic discovers the coupling wear upon guard disassembly. There are no safety issues if appropriate ppe is used upon pump cleaning before repair. An environmental alert is used and all operation within the premises are halted until the pump is put back to operate. Operation in the whole area stops for 1 hour until work permits are ready and coupling is replaced. This represents a cost of about $500,000.00. Repair costs amount about $100.00. There is a chance of losing raw material inside chemical reactor if the repair takes more than 2 hours resulting in additional $300,000.00 cost. Misalignment may cause a 10+% increase in electrical consumption. This failure has occurred several times. The life of the coupling is reduced to just 6,000 hours if not properly aligned.	479	P
					2	Pump cavitation	MEC a	Low suction pressure due to a dirty strainer	AGE	A low pressure condition is indicated in the HMI. There are no safety issues if appropriate ppe is used upon strainer cleaning and replace which takes about 10 minutes. There are no environmental issues. Production delay is not significant (about $1,000.00 in deferred production). The labor and material cost for the task is about $50.00. No secondary damages are associated with the failure. It has occurred several times and it is likely to happen passed 300 hours of operation.	3.2	P

*What happens when the failure occurs?
1. What are the facts proving the failure occurrence?
What would happen if a multiple failure, in the particular case of a hidden failure, takes place?
2. How is the safety of the people around the asset affected?
3. How are environmental goals impacted?
4. How is production or the operation affected by the failure?
5. What kind of physical damage is caused by the failure?
How costly is the failure in terms of maintenance and repair?
6. Are there secondary damages? What can be done to restore operations? How long would it take?
7. How likely is the failure to occur? Has it happened before?

**
H - Hidden
S - Safety or environmental
P - Production
M - Maintenance

FIGURE 10.2
RCM-R® Caustic soda pump consequence selection example.

RCM-R® helps practitioners to find the right consequence management task through the use of a decision algorithm diagram. Let's begin with a look at some typical RCM decision diagrams used today in practice for each of the four consequence management classifications. The starting point in the decision diagram is precisely the failure consequence type for each failure cause. The analysis team, using their expertise, must find specific tasks and appropriate task frequencies to properly mitigate failure consequences.

HIDDEN FAILURES CONSEQUENCE MANAGEMENT TASKS

Experience tells us that most hidden failures are best treated with type D or detection tasks aimed at identifying defective protective devices or redundant components. In fact, over 30% of the tasks recommended in modern manufacturing machinery are based on go/no go verification of protective devices, redundant components, and condition monitoring instruments. However, if the failure poses serious safety or environmental risks, we should avoid waiting until it happens. Thus, typical RCM decision diagrams require that we investigate whether any proactive task is technically feasible and worth doing to avoid the failure. Condition monitoring tasks are always considered first in the decision diagram, due to the fact that they are usually not intrusive, allowing the asset to continue to operate while maintenance or operation personnel perform them, and because statistically random failures can be expected to arise most often. A typical (SAE JA1011 compliant) approach to tackle hidden failure consequences is shown in Figure 10.3. Note that the diagram suggests a redesign task may or must be considered to avoid the failure, since the event of a multiple failure may become catastrophic to the integrity of the machine or its operators.

SAFETY AND/OR ENVIRONMENTAL CONSEQUENCE MANAGEMENT TASKS

Safety risks may expose asset operators, maintainers, or even the public in general to accidents affecting their physical integrity or even causing

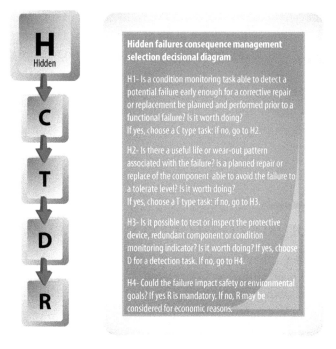

FIGURE 10.3
Typical hidden failure decision diagram.

death. Safety (and/or environmental) consequence management processes must ensure that the identified risks are eliminated or reduced to the minimum probability of occurrence or that consequences are mitigated to a level that is considered tolerable. The sequence of exploring task possibilities is the same as it was for hidden failures. Condition monitoring is the first consequence management policy considered. Timed restoration or replacements are the second choice for this type of failure consequence. In the case where condition monitoring or time-based tasks alone are not able to reduce the probability of failure occurrence to tolerable levels, then a combination of C and T tasks can be considered. Thus, you can combine monitoring the condition of a particular component with replacing a second one in the same task if life data analysis tells you that the failure of the second component exhibits a strong wear-out pattern. Sometimes, no proactive task or combination of proactive tasks is found to be capable of reducing the likelihood of the failure to an acceptable rate of occurrence. Then, a redesign task must be carried out to make the likelihood of failure acceptable to the owner of the asset. Figure 10.4 shows a classic decision diagram concerning type S consequence failures. Note that there

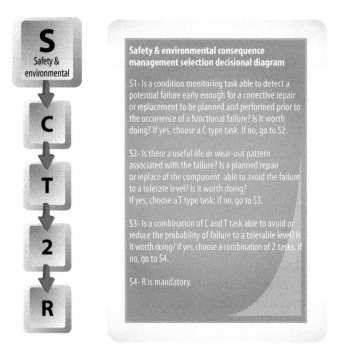

FIGURE 10.4

Typical safety and environmental consequence failure decision diagram.

is no room for F (run to failure) for critical safety or environmental consequence types of failure. Most of the time, the analysis team identifies the risk, and further evaluation is needed to produce a final solution (redesign task) to reduce the failure likelihood to tolerable levels. The authors have seen cases that have been referred to the asset's designer or manufacturer for further evaluation and eventual recommendation of an acceptable solution to handle safety risks.

PRODUCTION CONSEQUENCE MANAGEMENT TASKS

Evident failures not impacting safety and environmental goals are treated slightly differently from the previous two consequence types. C-type maintenance tasks are also highly preferred, because they seldom interfere with production activities, as opposed to T tasks, which normally require machines to be stopped for their execution. The main difference in the failure management policy with P-type consequence failures is that run to failure may be allowed under some circumstances. That is, failures for which it can be demonstrated that maintenance efforts are costlier than the production losses attributed to the failure event are normally allowed to occur. This is true for both production and maintenance types of consequence failures. This can be demonstrated in the failure effects statement by comparing the cost of an actual proactive task with the expenses incurred in letting the failure happen. There may also be technical reasons related to the physics of the failure. For instance, some failures may be very difficult to predict or prevent using technologies available at the job site. Some people say "you can't prevent what is not preventable nor can you predict what is not predictable." Technically speaking, that is what we check for when we make sure that tasks are always "technically feasible" or viable. A CM-R® training participant recently told one of the authors that he eliminated a lot of T type tasks after his team determined they were costlier to carry out than the production losses that their failures caused. Those decisions were changed to F (run to failure). His company began saving money on consumables no longer required for preventive replacements, but more importantly, they also avoided some maintenance-induced failures. In the past, they had experienced premature failure of replaced components after the replacement had been carried out. Finally, the analysis team may recommend redesign tasks for failures with type P consequence if those consequences are deemed too costly. The intention of R recommendations is to reduce the failure likelihood or its consequences to a level tolerable to the user. Figure 10.5 shows a classic decision diagram concerning type P consequence failures.

MAINTENANCE CONSEQUENCE MANAGEMENT TASKS

There are failures that do not impact production, safety, or environmental goals. They are relevant only because of the high repair costs they would incur should they happen. A bottling company with four centrifugal air compressors lost one due to an operational mistake. The 250 HP compressor replacement cost was over 125,000 USD. But, production was not affected, because the plant air demand required just three of the four available units. Would it really matter to the company owner if the 125,000 USD loss was due to production losses or to a failed filler machine? Of course not—the company loses the same amount either way. Therefore, failures impacting

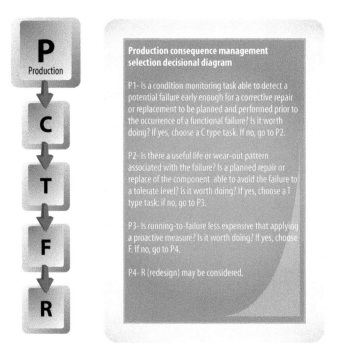

FIGURE 10.5
Typical production consequence failure decision diagram.

maintenance costs may be as critical as those affecting production capacity, as long as their repair cost is sufficiently high. In fact, typical RCM consequence management decision diagrams offer similar maintenance policies priority sequences in their decision diagrams for both type P and M consequence failures. Consider, for instance, the high costs a company is paying for the combined effect of compressed air leaks, extra energy costs, and reduced bearings life due to shaft misalignment and rotor imbalance. Maintenance-related costs can add up significantly. But in many cases, they are overshadowed by production-related losses. Figure 10.6 shows a classic decision diagram concerning type M consequence failures. Note that it follows exactly the same sequence of consequence management policies as for production-type consequence failures.

A simplified combined diagram summarizing all the processes described in Figures 10.3 through 10.6 is presented below as Figure 10.7. Failure effects consequences codes are found in the first column. The right consequence is selected with the help of Figure 10.1. Then, moving from left to right, a consequence management policy can be selected, as they have been placed in the order of priority recommended by the previously

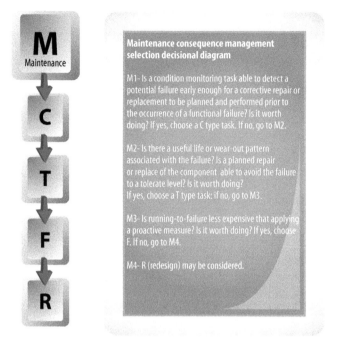

Maintenance consequence management selection decisional diagram

M1- Is a condition monitoring task able to detect a potential failure early enough for a corrective repair or replacement to be planned and performed prior to the occurrence of a functional failure? Is it worth doing? If yes, choose a C type task. If no, go to M2.

M2- Is there a useful life or wear-out pattern associated with the failure? Is a planned repair or replace of the component able to avoid the failure to a tolerate level? Is it worth doing? If yes, choose a T type task: if no, go to M3.

M3- Is running-to-failure less expensive that applying a proactive measure? Is it worth doing? If yes, choose F. If no, go to M4.

M4- R (redesign) may be considered.

FIGURE 10.6
Typical maintenance consequence failure decision diagram.

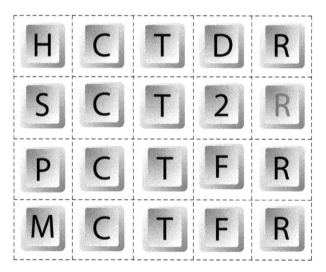

FIGURE 10.7
Simplified consequence failure decision diagram.

shown individual diagrams. For instance, if a production consequence is selected, the first consequence management choice would be condition monitoring, followed by time-based maintenance, run to failure, and redesign.

RCM-R® FAILURE CONSEQUENCE MANAGEMENT DECISION DIAGRAM

The process for failure consequence management selection according to the SAE JA 1011 standard "Evaluation Criteria for Reliability-Centered Maintenance (RCM) Processes" was discussed back in Chapter 3. It is one of the most important aspects of the standard, leading to the optimization of maintenance effectiveness for critical assets as experienced by the aviation industry back in the 1970s. RCM-R® uses Figure 10.1 to identify the appropriate failure consequence according to the evidence of each failure cause. Then, the analysis team answers a series of nine questions, leading them to select the most appropriate consequence management policy for each failure cause. The questions are organized in the form of a decision diagram, as shown in Figure 10.8.

The analysis group facilitator must ensure that the team understands what is meant by each of the questions. Many RCM practitioners today do not use a team and decision diagram approach. Alternatively, they use "RCM" software relying on the input of only one person, usually from maintenance, and not necessarily having a complete understanding of the system's operation or even of maintenance management concepts. The result of this is often the emergence of faulty justification of the current time-based task-driven maintenance plan. Of course, when they get to that conclusion, they further conclude (correctly) that their effort was largely wasted, because nothing new came from it. But their approach was flawed. We are proponents of the trained team-based approach and the use of a decision diagram to avoid those sorts of errors. Let's see what each of the nine questions in the RCM-R® decision diagram is looking for.

1. *Is there a significant safety, environmental or economic consequence?*
   All failure causes must have been evaluated for their impact on business goals at this stage of an RCM-R® analysis—those evaluations appear in the descriptions of effects. F (run to failure) can be

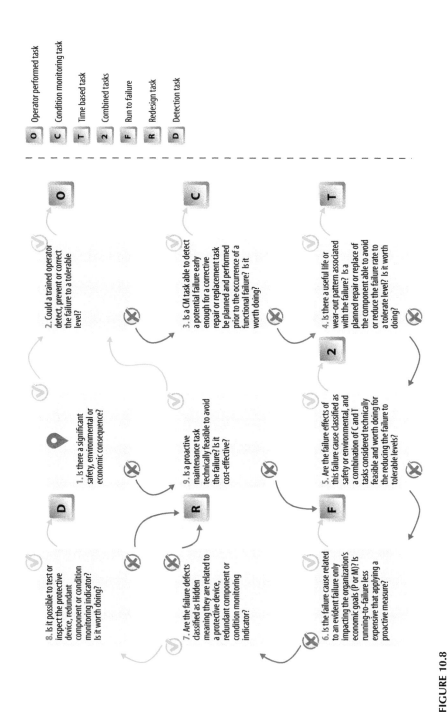

FIGURE 10.8
RCM-R® Failure consequence management decision diagram.

accepted only if a failure is considered insignificant. Therefore, we must separate critical from noncritical failures at this time. This is done through the RCM-R® failure effects risk evaluation matrix, shown in Figure 7.4. The risk number obtained as the result of the analysis and shown in Figure 7.5 will tell the analysis team whether the failure cause is significant or not. Significant failures continue their evaluation with Question 2, while insignificant failure causes take a different path to Question 9.

2. *Could a trained operator detect, prevent, or correct the failure to a tolerable level?*

Operators play a significant role in asset management. They can perform many proactive tasks while doing their routine plant operation activities. Total productive maintenance (TPM) considers operators performing asset cleaning, lubrication, adjustment, inspection, and minor maintenance to avoid asset deterioration, keeping the asset in proper operating condition. Operators are closer to the operating assets and can respond immediately to perform routine activities without having to wait for maintenance technicians. Of course, they must be properly trained in recognizing the early signs of asset deterioration and in the execution of the associated corrective tasks to solve them. An organization implementing TPM benefits from this by getting minor maintenance tasks almost free of charge, performed by otherwise idle operators, while freeing up maintainers to use their specialized skills on more complex maintenance. Time-based tasks such as replacement of consumables such as filters and lubricants are examples of what a trained operator can do. Some other detection and condition monitoring, visual inspection, and safety check tasks can be carried out by qualified operations personnel. O tasks need to be included in formal PM work orders and assigned to operators by the maintenance planners, as is done with maintenance craft PMs. An effective PM program including O tasks requires that operation supervisors be accountable for their execution. Remember that these tasks were the result of multidisciplinary team analysis agreeing on them with the participation of production representatives. O tasks are selected when it is found that trained operators are able to detect, prevent, or correct the failure cause. When a particular failure cause cannot be treated effectively with an O task, we should go further to evaluate it with Question 3. An organization not willing to use O tasks can use the decision diagram depicted in Figure 10.9.

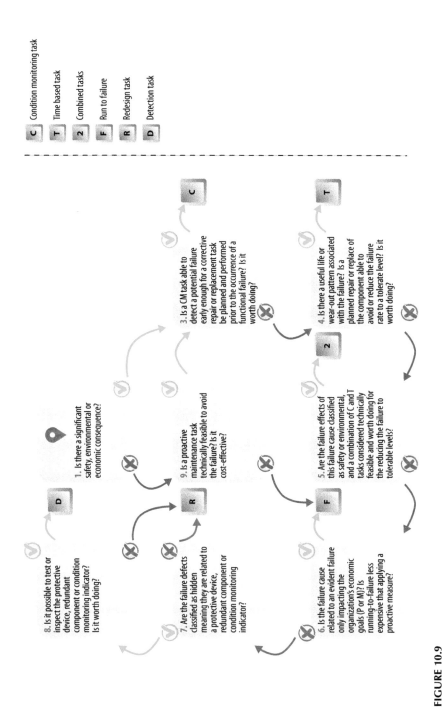

FIGURE 10.9

4. RCM-R® Failure consequence management decision diagram with no O tasks.

3. *Is a condition monitoring task able to detect a potential failure early enough for a corrective repair or replacement task to be planned and performed prior to the occurrence of a functional failure? Is it worth doing?*

Condition monitoring has been explained extensively, particularly in Chapter 8. It is important that the analysis team understands that condition monitoring entails measuring, trending, and analyzing asset health indicators over time. Most of the C tasks do not require the use of complex predictive maintenance technologies and can be done by trained operators or maintainers using the human senses or through monitoring process operating and performance parameters. Many tasks comprise analyzing process parameters, such as temperature, flow, load, and fluid levels, and so on. Other tasks may require measurement of torque for bolts, voltage, amps, load readings, and so on. There is also a need to use PM and NDT for detecting critical potential failures. The analysis team should proceed to Question 4 if a C task cannot be found to avoid the failure or reduce its occurrence to satisfactory levels.

4. *Is there a useful life or wear-out pattern associated with the failure? Is a planned repair or replacement of the component able to avoid the failure or reduce its probability to a tolerable level? Is it worth doing?*

The opinion of experts supported by failure event data is instrumental for confirming age-related failures. Beware of a natural tendency to overstate the significance of some failures. For example, a team member from operations may well remember some premature failure events of particular components. This may lead the team to consider the wrong failure pattern for the component under normal operation. It is possible that the team is not considering that there are dozens of nonfailed components, and that the two failures mentioned were due to unusual circumstances. Premature failures can happen after any intervention, and they can happen to components that fail randomly as well as those that fail with age or usage. There is always a risk that the physics of the component failure may be misunderstood due to the consideration of only outlier events. T tasks are recommended when the failure events analysis confirms a wear-out pattern, and the cost effectiveness of the preventive tasks is ratified. Otherwise, the team should proceed to Question 5.

5. *Are the failure effects of this failure cause classified as safety or environmental, and is a combination of C and T tasks considered technically feasible and worth doing to reduce the failure to tolerable levels?*

This question is intended to separate failures with safety or environmental consequences from those that impact only the economy of the organization. A combination of C and T tasks can be considered if single tasks alone cannot avoid the failure or reduce the failure rate to tolerable levels. For such cases, the two (combination of tasks) consequence management policy is adopted. Note that this is only feasible if the failure mode cause is age or usage related, and there is a condition that can be monitored. The analysis team should go to Question 6 if the combination of tasks for a safety or environment consequence is unable to avoid the failure or reduce its likelihood to tolerable levels. They should also go straight to the following question if the failure consequence is not classified as S.

6. *Is the failure cause related to an evident failure impacting only the organization's economic goals (P or M)? Is running to failure less expensive than applying a proactive measure?*

This question looks for evident failures causing impact only on economic goals. In other words, we are looking for P- or M-type consequence failure causes. These failures are considered and treated in a different way from S- or H-type consequence failure causes. In the case of these P or M failure causes, we will let them run to failure if no proactive maintenance tasks are found to be cost effective. Therefore, an affirmative answer to Question 6 will automatically assign an F consequence management policy for the failure cause under consideration. The analysis team should proceed to Question 7 if running to failure is considered unacceptable or if the failure is not evident.

7. *Are the failure defects classified as hidden (meaning that they are related to a protective device, redundant component, or condition monitoring indicator)?*

Hidden failures are filtered out and treated at this stage by answering this question. Figure 10.7 shows that all failure effects consequence types typically follow the same priority order, consisting of C followed by T consequence management policies up to the second stage of the analysis. Then, three different paths, determined by the evidence of the failure, are possible to follow. Evident failures with safety consequences follow one path, whereas evident failures with merely economic consequences follow another particular path. Finally, there is a unique path for hidden failures. In Chapter 3, we point out that SAE JA1011 requires that hidden failures be highlighted and separated from evident failures. The later type of failures must also be classified as to their impact on safety or economic goals per SAE JA1011. RCM-R® clearly satisfies and goes beyond this requirement. If the answer to this question is affirmative, then the analysis team can go straight to the next question. Otherwise, R may be required. S-type consequences require mandatory R actions to deal with the failure consequences. R implies modifications of the asset itself, its operation or maintenance procedures, or some other factor (e.g., training). R actions for P- or M-type consequence are optional and may be considered.

8. *Is it possible to test or inspect the protective device, redundant component, or condition monitoring indicator? Is it worth doing?*

Hidden failures that reach this stage of consequence management analysis are undoubtedly critical. Therefore, redesign should be considered if there is no way to test their functionality for hidden failures. Remember that failure causes have already been evaluated for the possibility of detecting potential failures or avoiding functional losses due to age or usage by this stage. The type of redesign tasks would depend on the severity of the possible damages caused by the failure. Could there be personal injuries, or just production loss, or only maintenance-related expenses? The answer to this question would drive the course of action regarding the type of redesign task or one-time change needed to tackle the risks. Detection tasks are then assigned when testing protective devices, redundant components, or condition monitoring instruments is technically feasible and cost effective.

9. *Is a proactive maintenance task technically feasible to avoid the failure? Is it cost effective?*

We determined what to do when Question 1 was answered affirmatively. In this case, we started answering Questions 2 through 8 in an orderly fashion. Now, we are considering the case in which failures are considered noncritical. The straightforward consequence management treatment for them would be F (run to failure). This question offers a second chance to think about a possible proactive measure being considerably less costly than letting the asset run to failure. Consider the case of an analysis on a very expensive and critical asset in the oil and gas industry. The production representative on the analysis team established that failures causing a production loss of less than X amount of money were regarded as noncritical provided they did not impact safety or environmental goals. The X threshold for determining criticality was in the range of six figures USD. In this case, some 20 failure causes were in the vicinity of that figure. They could have been considered for consequence management policy F, but combined, they amounted to several million USD. They were evaluated by answering this question. Surprisingly, cost-effective O, C, T, and D tasks were found for some of them, representing a significant cost-saving opportunity for the organization. No tasks considered to be worth doing were found for others, so they were allowed to run to failure. Thus, F is considered if no cost-effective proactive task is found for a failure cause that poses only economic consequences. The analysis team should go on to answer Question 2 if they understand that a cost-effective task may be found to avoid the failure. It is the team's call to consider skipping Question 9, provided that the failure causes were already considered noncritical in the preceding questions, getting to F without further consideration.

The RCM-R® consequence management decision diagram has been converted into a simplified flowchart with coded boxes, as shown in Figure 10.10. The flowchart guides the analysis group through the questions based on the answers to the previous questions. To avoid mistakes, the analysis team should have the nine key questions available for reference when using this chart. A yes answer to Question 1 would take you to Question 2, a no answer to Question 9, and so on.

DEVELOPING MEANINGFUL MAINTENANCE TASKS

The RCM-R® process (and any other SAE-compliant process) will enable the analysis team to select appropriate failure consequence management policies that are both technically feasible and worth doing. We can use either the decision diagram, depicted in Figure 10.8, or its corresponding coded flowchart, as shown in Figure 10.10, to accomplish this endeavor. But, no RCM process by itself will render actual maintenance tasks from the failure consequence management policies on its own. The decisions

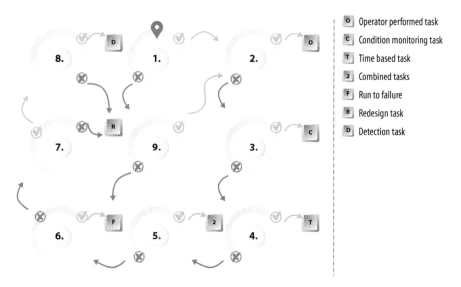

FIGURE 10.10
RCM-R® Simplified consequence management decisional flowchart.

define your failure consequence management policies, but you must still define the actual tasks to be executed in the field. Consequently, the analysis team must produce task descriptions as a part of the process. The tasks will eventually be performed by maintainers and operators. Ultimately, proactive (types C and T) and detective (type D) tasks are grouped by craft type and by execution frequencies in the form of PM work orders. Sometimes, these groupings can be converted into "routes" that maintainers or operators follow while executing tasks in a defined sequence, making the work more efficient to execute. Decisions result in tasks, and tasks end up in work orders. There is some general information needed at the work order level and some other specific detail necessary at the task level. Furthermore, some tasks may reference standard maintenance procedures (SMPs) to avoid unnecessary lengthy writing as part of the PM tasks. Some computerized maintenance work management systems have standard task definitions that are commonly used in building work orders.

Each PM work order must establish the personal protective equipment required for the job. All safety and environmental hazards faced while performing the job should be mentioned, as well as a complete list of tools, materials, and spare parts. The definition of skills and category of crew (i.e., industrial mechanic, certified electrician, instrumentation technician, etc.) needed to perform the PM work order must be clearly defined in the PM document. Each individual task within the PM document should be precise and concise, clearly establishing

- Who is doing the job: Skill requirement.
- What is to be done: A brief statement clearly establishing what exactly is being done (e.g., check and record vibration readings).
- How the work is done: What methodology, procedure, and specialized instrumentation are needed for the job (e.g., use the XYZ handheld overall vibration meter applied to the measuring points on the bearing housings)?
- Where the work is being specifically applied: At which specific location or place within an asset is the task is being performed (e.g., take vertical and horizontal measurements on pump P 123A at the motor inboard and outboard bearings, the pump inboard and outboard bearings, and an axial reading at the motor outboard bearing)?
- Time required to perform the task.
- Number of people required to perform the task.
- References to other documents needed to perform the job.
- If the task is a C task, include a statement of what to do if the condition monitoring reveals an incipient failure.

We should remember that tasks are aimed at failure root causes and should be able to avoid the failure or reduce its consequences to acceptable levels to the owner/operators of the asset. There is a direct link between the failure cause and the eventual tasks coming from the consequence management policies produced by the RCM-R® process. Complex tasks may require targeted documented standards or SMPs detailing the steps and describing how to perform them. For instance, a polarization index (PI) test may be required to address the failure cause "degradation of cable insulation due to age." Thus, the assigned task may read "Perform a PI test to motor following SPM 2304E Procedure for conducting PI test for AC induction motors." The SPM must establish all the details, including instrumentation, methodology, and PPE, for carrying out the task.

CHAPTER SUMMARY

RCM-R® classifies failure effects into four categories based on evidence of failure impact on safety, the environment, operational capability, and cost. P (production), M (maintenance), S (safety and environmental), and

H (hidden) are the four possible types of failure consequences the analysis team must choose from to classify every failure effect. Only one category must be chosen—the most severe. The failure effects classification questionnaire shown in Figure 10.1 helps the users to choose the failure consequence category best suiting each particular failure cause. The next step of the analysis embraces deciding which consequence management policy must be chosen for each critical failure cause according to its identified failure consequence classification: O (operator task), C (condition monitoring), T (time-based restoration or replacement), 2 (combination of two tasks), D (detection), F (run to failure), and R (redesign).

RCM-R® uses a consequence management decision diagram to select the most appropriate consequence management policy for every failure cause identified by the multidisciplinary analysis team. The decision diagram shown in Figure 10.8 guides the analysis through a process compliant with SAE JA1011, treating each failure cause according to its impact on the organization's goals (regarding safety, environmental, and economic goals) and also according to the evidence of failure as seen by operation and maintenance personnel during normal operation.

As an example, the diagram is used to assign consequence management policies to two failure causes, as shown in Figure 10.11. Note that each answer leads the team to its next question, till a consequence management policy is finally decided for application. The answer sequence for failure cause 2-A-1-c (coupling wear due to misalignment caused by wrong practices) in the example shown in Figure 10.11 is as follows:

Question 1: Answered Yes (the failure causes significant production loss)
Question 2: Answered No (meaning no operator task is suitable)
Question 3: Answered No (meaning no condition monitoring task is suitable)
Question 4: Answered No (meaning no time-based task is suitable)
Question 5: Answered No (the failure consequence is not affecting safety)
Question 6: Answered No (meaning the failure is evident, but F is not accepted)
Question 7: Answered No (meaning the failure is not hidden)

Then, R is selected as the consequence management policy. Therefore, a modification of maintenance practices is needed to avoid failure or reduce its rate to acceptable levels. Some redesign tasks may be defined

Failure modes/ types & root causes/ type — Task

#	Failure mode	Type	#	Root causes	Type	Consequence*	Consequence management policy**	E- Elect M- Mech I-Inst R - HVACR O-Ope	Task description	Time (hours)
1	Pump coupling wear	MAT	c	Misalignment caused by wrong practices	MGT	P	R	M	Prepare an aligment SMP & train & certify mechanics on industry best alignment practices	N/A
2	Pump cavita-tion	MEC	a	Low suction pressure due to a dirty strainer	AGE	P	O	O	Remove, clean with solvent X & replace the pump suction strainer	0.2

*	**
H - Hidden	O - Operator M
S - Safety or environmental	C - Condition M
P - Production	T - Time based
M - Maintenance	2 - Combined
	D - Detection
	F - Run to F
	R- Redesign

FIGURE 10.11

RCM-R® Caustic pump consequence management policy selection example.

later by other people with the required expertise, but to the analysis team, this looks like a clear case of ensuring that the maintenance practices are correct and applied.

The second failure cause evaluated was coded 2-A-2-a (pump cavitation due to low suction pressure caused by a dirty strainer) and was assessed as shown in Figure 10.8, as follows:

Question 1: Answered No (failure causes insignificant $1000 produc-tion loss)
Question 9: Answered Yes (task cost is 5% of production loss)
Question 2: Answered Yes (operator can perform a suitable task)

In this case, O is selected as the consequence management policy. Therefore, a task performed by operators was found to be cost effective to avoid the failure. Figure 10.12 indicates the two tasks assigned to the failure causes considered in the caustic soda pump example.

	Failure modes/ types & root causes/ type						Task		
#	Failure mode	Type	#	Root causes	Type	Consequence*	Consequence management policy**	E- Elect M- Mech I-Inst R - HVACR O-Ope	Task description	Time (hours)
1	Pump coupling wear	MAT	c	Misalignment caused by wrong practices	MGT	P	R	M	Prepare an aligment SMP & train & certify mechanics on industry best alignment practices	N/A
2	Pump cavita-tion	MEC	a	Low suction pressure due to a dirty strainer	AGE	P	O	O	Remove, clean with solvent X & replace the pump suction strainer	0.2

*	**
H - Hidden	O - Operator M
S - Safety or	C - Condition M
environmental	T - Time based
P - Production	2 - Combined
M - Maintenance	D - Detection
	F - Run to F
	R- Redesign

FIGURE 10.12

RCM-R® Caustic pump maintenance tasks example.

11

Fine-Tuning RCM Analysis

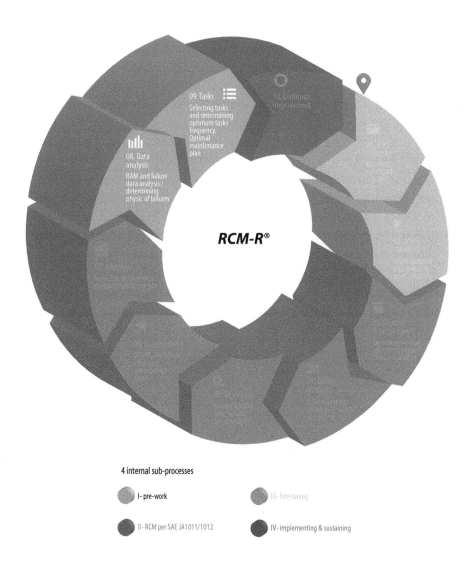

09. Tasks
Selecting tasks and determining optimum tasks frequency. Optimal maintenance plan

10. Continual improvement

08. Data analysis
RAM and failure data analysis / determining physic of failures

RCM-R®

4 internal sub-processes

I- pre-work

III- fine tuning

II- RCM per SAE JA1011/1012

IV- implementing & sustaining

THE NEED FOR BETTER DATA

Asset data integrity was identified in Chapter 3 as one of the five pillars of RCM-R® together with RCM, reliability, availability, and maintainability (RAM), Weibull analysis, and continuous improvement. In general, RCM-R® requires operational, technical, reliability, maintenance-related, failure, material, financial, safety, and environmental data, which is analyzed for decision-making purposes. Maintenance- and operational-related documentation on downtime, spare parts consumption, total preventive maintenance (PM) man-hours, people skills, corrective maintenance man-hours, failure events, quality defects, etc. are needed. We saw in Chapter 3 that the premise of the RCM-R® process is that failure data documented in corrective work orders can be statistically analyzed to find the predominant failure pattern of each critical failure cause. With this, an RCM-R® analysis can be fine-tuned using statistical failure data analysis to develop better maintenance strategies and tasks interval assignment. Work order data is the framework of any good reliability improvement program. Therefore, maintenance and reliability engineers must make sure that important failure and repair data is included in their critical assets corrective and proactive work orders. If we apply the process shown in Figure 4.1 (creating value from asset data) to good work order data and process it with some reliability analysis methods, we will obtain the benefit of an improved and evidence-based RCM-R® decision process. SAE JA1011 requires that "any mathematical and statistical formulae used in the application of the process (especially those used to compute the intervals of any tasks) be logically supportable, available to and approved by the owner or user of the asset." This fine-tuning of our RCM-R® analysis meets that requirement.

RELIABILITY, AVAILABILITY, AND MAINTAINABILITY (RAM) ANALYSIS

Let's define some basic terms used in reliability engineering as the starting point of this quantitative analysis discussion.

- Reliability: The ability of an item to perform a required function under stated conditions for a stated period of time.

- Reliability (as a probability): The probability that an item will perform a required function under stated conditions for a stated duration of operating life.
- Corrective maintenance: An unscheduled activity performed as a result of a failure to restore an item to a specified level of performance.
- Proactive maintenance: A scheduled inspection, detection, repair, or replacement task for retaining an item at a specified level of performance.
- Proactive maintenance frequency (Fp): The frequency at which proactive maintenance tasks are performed. It is expressed mathematically as the inverse of the time between planned proactive tasks.
- Mean time to failure (MTTF): The average of the observed age at failure of similar nonrepairable items.
- Mean time between failures (MTBF): The average of the observed age at failure of similar repairable items.
- Constant failure rate (λ): The number of failures per unit time.
- Mean time between maintenance (MTBM): The average time between planned and unplanned maintenance work applied to a particular asset or component.
- Mean corrective time (Mct): The mean time corrective work takes from failure identification to total recovery from a functional loss.
- Mean maintenance time (M): The average time to conduct all maintenance work taking into consideration both planned and unplanned interventions.
- Mean down time (MDT): The average time an asset is not operating due to maintenance and/or other causes.
- Inherent availability (Ai): The proportion of time an asset is available to operate due to failure only.
- Achieved availability (Aa): Proportion of time an asset is available to operate due to maintenance (both corrective and proactive). Aa < Ai.
- Operational availability (Ao): Proportion of time an asset is available to operate due to the influence of maintenance (both corrective and preventive) and other activities causing operation to cease, such as personnel meetings, holidays, lack of raw materials, etc. Ao < Aa < Ai.

MTBF is considered a reliability parameter in RAM analysis. By the same token, Mct, often known as mean time to failure (MTTF), is the quantitative parameter related to maintainability, and it is measured in repair time. Maintainability is a design characteristic of assets dealing with ease, accuracy, safety, and economy in the execution of maintenance functions. Equations 11.1 and 11.2 express how MTBF and Mct are calculated, respectively. It is observed from Equation 11.3 that both reliability and maintainability influence the Ai of assets. Many companies today define Ai expectations for plant systems. Then, plant designers must take this information into consideration when selecting systems configurations enabling proper design characteristics yielding the necessary reliability and maintainability performance for the required Ai.

$$MTBF = \frac{\text{operating time}}{\text{number of failures}} \tag{11.1}$$

$$Mct = \frac{\text{total failure downtime}}{\text{number of failures}} \tag{11.2}$$

Sometimes, we are asked about the availability of a particular asset without realizing that there are basically three ways of calculating it. If you are a corrective maintenance supervisor, you would be interested in calculating Ai to anticipate your workload or to plan improvements of your corrective maintenance check lists to reduce the active repair time. Ai is defined mathematically in Equation 11.3.

$$Ai = \frac{MTBF}{MTBF + Mct} \tag{11.3}$$

For instance, if our caustic water pump experienced five failures in a period of 10,000 operating hours when applying Equation 11.1, its resulting MTBF would be 2,000 h between failures. This pump is part of a greater air scrubber system having a production room fugitive emissions exhaust fan. Similarly, if our fan exhibited just two failures in the same number of operating hours, its MTBF would be 2.5 times greater (5000 h) than that observed for the pump. Can we consider the pump as a bad actor due to the fact that it is failing more often than the system's exhaust fan? Certainly, the pump is showing lower reliability, and the maintenance supervisor should look for ways to improve its repairs, perhaps through the use of precision maintenance techniques. Let's assume that fan repairs take about 72 h on average due to inaccessibility issues and the need for isolation for handling safety and environmental risk. Which of the two components would have more effect on the scrubber system availability

TABLE 11.1

Inherent Availability Calculation Results

Asset ID	Operating Time (h)	Number of Failures	Failure Downtime (h)	MTBF (h)	Mct (h)	Ai
Caustic water pump	10,000	5	10	2,000	2	0.9990
Scrubber system fan	10,000	2	144	5,000	72	0.9858
Scrubber system	10,000	7	154	1,429	22	0.9848

if pump repairs only took 2 h on average? See the calculated Ai for both pump and fan and also for the system as a whole in Table 11.1 with the use of Equation 11.3.

The scrubber fan certainly has better reliability than the caustic pump. But, since it requires more repair time, its Ai is lower than that shown by the pump. The scrubber system's availability can also be calculated by taking into account all failures for the pump and exhaust fan in conjunction with the total downtime caused by the failures of the two components.

Proactive maintenance work that interferes with production activities also affects asset availability. The maximum possible availability an asset can attain when influenced by both proactive and corrective maintenance work is Aa. Theoretically, it is the maximum proportion of time the equipment would be available to produce if maintenance activities were the only ones causing production delays. Consider, for instance, our caustic water pump and exhaust fan having a PM program requiring 2 h of preventive work interventions for each component every 1000 operating hours. The production supervisor would be unable to use the scrubber during that time, further affecting the system availability. The actual system availability experienced by the system owner due to the influence of both corrective and proactive maintenance activities is calculated as Aa. Some other calculations are needed to get the actual Aa of the pump, as it is influenced by the combination of both corrective and proactive maintenance. See the formulae used for calculating Aa below. Let's define the formulae for calculating the failure rate, provided it is constant, by the use of (11.4).

$$\lambda = \frac{1}{\text{MTBF}} \tag{11.4}$$

Then, let's show how to determine M considering that both corrective and proactive work will stop operational activities. Preventive maintenance

(PM) frequencies are also expressed in terms of activities taking place per operating time. For example, a PM frequency carried out every 1000 operational hours is expressed as 1/1000, meaning that a PM intervention is taking place every 1000 operating hours. We will name the variable Fpt as the PM frequency. Then,

$$Fpt = \frac{1}{\text{Time between PMs}} \qquad (11.5)$$

We need to calculate two more variables to determine Aa, and they are M and MTBM. Let's mathematically define M first as

$$M = \frac{Mct \times \lambda + Mpt \times Fpt}{\lambda + Fpt} \qquad (11.6)$$

Then, MTBM can be calculated with the following mathematical expression:

$$MTBM = \frac{1}{\lambda + Fpt} \qquad (11.7)$$

Aa is then calculated after finding the value of both MTBM and M as follows:

$$Aa = \frac{MTBM}{MTBM + M} \qquad (11.8)$$

Now, we can apply Equations 11.6 through 11.8 to calculate the pump's Aa as shown in Table 11.2. Note that all availabilities drop slightly for both individual components and the system as a whole due to the influence of the PM work.

Asset operation is not halted by maintenance activities alone. Sometimes, the lack of raw materials, operating and maintenance personnel meetings,

TABLE 11.2

Achieved Availability Calculation Results

Asset ID	MTBF[a] (h)	λ (1/ MTBF)	Fpt (1/h)	Mct[a] (h)	Mpt (h)	M (h)	MTBM (h)	Ao
Caustic water pump	2000	0.0005	1/1000	2	2	2	667	0.9970
Scrubber system fan	5000	0.0002	1/1000	72	2	13.7	833	0.9839
Scrubber system	1429	0.0007	1/1000	22	4	11.4	588	0.9810

[a] From Table 11.1.

holidays, and other activities curtail production time, affecting the assets' Ao for production. Ao is affected by both maintenance and nonmaintenance activities that reduce physical assets' operation time. If we considered the scrubber system's MDT to be 25 h due to all causes, then the system would yield an operational availability of 0.9592 using Equation 11.9, which represents a drop of nearly 2.25% of its available time as compared with its corresponding Aa.

$$Ao = \frac{MTBM}{MTBM + MDT} \qquad (11.9)$$

RAM analysis is an essential tool of RCM-R®, enabling the analysis team to define assets' reliability, maintainability, and resulting availability under their current operating context. Actions to improve maintainability or reliability can be taken if the current system's availability is not acceptable to its owner. Vital information is drawn from both corrective and proactive work orders for determining these important quantitative parameters. The need for better data becomes a real issue when a company wishes to improve its operational yield by optimizing the operation's availability for increased profitability. Keeping RAM parameters as key performance indicators (KPIs) is not difficult once appropriate data is there in the work orders for its calculation. RAM analysis is quite easy to calculate. RAM analysis is also flexible, as it can be applied to a single asset or to a whole system by simply including in the mathematics all failure and PM events of the desired system over a defined period of time.

RAM analysis has some limitations, being based on average data. Both repair times and time between failures calculations yield average data that is fit for the purpose of the analysis. Analyzing average data alone could, however, be misleading. The use of averages may mask the actual predominant failure patterns and lead to misapplication of consequence management policies (e.g., the use of averages could lead us to consider infant mortality failures as if they were random and having a low failure rate).

FAILURE DATA ANALYSIS

Statistical life data analysis complements RAM analysis by determining the physics of each failure cause and its corresponding characteristic life.

Reliability is seen as a probability of fulfilling a specified function instead of an average time to failure of an item, as is the case in RAM analysis. The analysis is particularly applicable to assets with operating and maintenance history with well-documented failure events. Such events should be recorded and sorted by failure causes. Some of the outcomes and applications of life data analysis according to Dr. Robert B. Abernethy, as he mentions in his publication *The New Weibull Handbook*,[1] are the following:

- Failure forecasting and prediction
- Evaluating corrective action plans
- Test demonstration for new designs with minimum cost
- Maintenance planning and cost-effective replacement strategies
- Spare parts forecasting
- Warranty analysis and support cost predictions
- Controlling production processes
- Calibration of complex design systems
- Recommendations to management in response to service problems

Mechanical, electrical, electronic, material, and even human failures can be modeled and predicted using failure data analysis techniques, as can other deficiencies related to quality control and design issues. We will discuss how failure data analysis can be very useful to complement the other pillars of RCM-R®. We will focus on the following aspects of failure data analysis:

- Creating a failure probability plot
- Determining reliability and probability of failure at any operating time
- Determining the item's predominant failure patterns (physics of the failure)
- Confirming appropriate consequence management strategies selection
- Calculating time-based task frequencies

We will also present tools for estimating type C tasks' optimum frequencies. This is important, because overmaintenance (resulting in excessive monitoring costs) is also possible if the wrong maintenance frequency is used in condition monitoring tasks.

WEIBULL ANALYSIS

Waloddi Weibull invented his Weibull distribution in 1937, claiming that it was useful for a wide range of problems and applications. The distribution, rejected by the statisticians of that era, later became the leading method in the world for analyzing life data. Later on, some other respected statisticians, such as Leonard Johnson from General Motors, helped improve on Weibull's plotting methods. Basic Weibull analysis consists of plotting failure data on Weibull probabilistic paper and interpreting the plot. Weibull plots are found to be very effective with extremely small samples of data for engineering analysis of even two or three data points. Predictions of failures and their corresponding costs, spare parts consumption, labor usage, failure rates, and electrical outages can be determined accurately through the use of this magnificent statistical tool.

We will discuss how to use and interpret four forms of the Weibull distributions, representing reliability, probability of failure, failure rate, and probability density function, for their practical use in failure data analysis. Let's start with the basic reliability function R(t) corresponding to the probability that an item survives to any given age. Letting T represent the time to failure and t the operating time, R(t) is the probability that the failure does not occur in the interval o to t. Then,

$$R(t) = e^{-\left(\frac{t}{\eta}\right)^{\beta}}$$

(11.10)

Figure 11.1 shows a reliability function plot of an item exhibiting a wear-out type of failure pattern. Note that the reliability of the item is considered to be 100% at the beginning of its operating life. Then, it continues to decrease till it reaches 0% reliability.

Let's take a look at the probability of failure or cumulative distribution function F(t) now. F(t) represents the probability of failure at or before operating age t. Then,

$$F(t) = 1 - e^{-\left(\frac{t}{\eta}\right)^{\beta}}$$

(11.11)

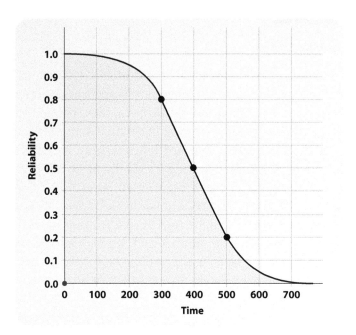

FIGURE 11.1
Reliability function plot.

Figure 11.2 shows a cumulative distribution function plot of an item exhibiting a wear-out type of failure pattern. Note that the unreliability of the item is considered to be 0% at the beginning of its operating life. Then, it continues to increase till it reaches 100% of probability of failure. Also, notice that

$$R(t) + F(t) = 1 \qquad (11.12)$$

The hazard function or failure rate is mathematically the first derivative of the function F(t). Then, its function is mathematically expressed as follows:

$$h(t) = \frac{\beta t^{(\beta-1)}}{\eta^{\beta}} \qquad (11.13)$$

The hazard function shows how the failure rate is changing with respect to time. An item experiencing random failures will show a flat horizontal failure rate line. Premature failures are portrayed with rapidly descending failure rate plots, while wear-out failures show an increased failure rate as

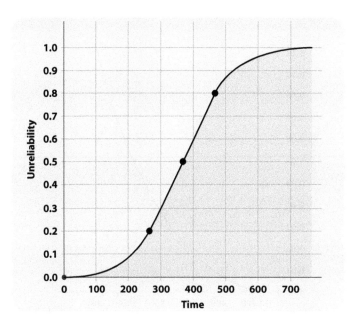

FIGURE 11.2
Cumulative distribution function plot.

time passes. Figure 11.3 showcases a hazard distribution function plot of an item exhibiting a wear-out type of failure pattern.

The area under the curve between two given ages of the probability density function (pdf) represents the probability that a new item fails at that given age interval. The probability density function can take many different shapes depending on the values of the Weibull parameter, particularly that of the shape parameter β. The pdf function becomes a normal distribution when

β equals 3.35. f(t) is mathematically expressed as follows:

$$f(t) = e^{-\left(\frac{t}{\eta}\right)^{\beta}} x \frac{\beta t^{(\beta-1)}}{\eta^{\beta}} \tag{11.14}$$

Figure 11.4 represents the distribution function exhibiting a strong wear-out failure pattern.

Creating and Interpreting Weibull Data Plots

The major advantage of Weibull analysis is its ability to provide accurate failure analysis and forecasts with extremely small samples. Our asset

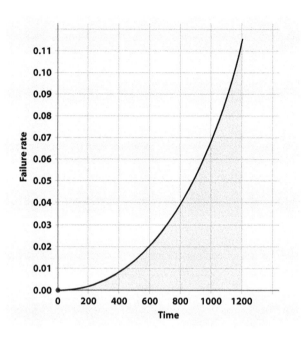

FIGURE 11.3
Hazard function plot.

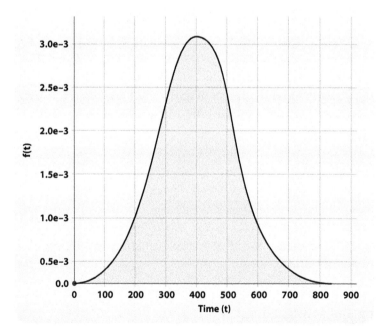

FIGURE 11.4
Probability density function plot.

management efforts to stem incidents of critical machinery failing can benefit from such analysis, which reveals the nature of the failure patterns being experienced. Predominant failure patterns, failure probabilities, consequence management policies, and optimum replacement times can be easily determined at the failure cause level. Another advantage of the method is that it provides a simple graphical plot of the failure data, which can be easily interpreted, somewhat intuitively, without the need for any calculation.

Weibull data plotting entails the graphing of failure time versus probability of failure on a particularly designed logarithmically scaled Weibull probabilistic paper. Therefore, it is a log plot of F(t), for which the horizontal scale of the plot is a measure of life or aging by the use of a time parameter (t). Life data means that we need to know the age of the items failing and in service. The time parameter t can be expressed in mileage (for vehicles), operating time, operating cycles, starts and stops, landings, takeoffs, storage time, etc. The best aging parameter is the one with the best fit compared with a straight line in the Weibull plot. There are two types of life data used for Weibull analysis: *standard life data*, consisting of exact failure age data, and *grouped or interval life data*, for which exact ages are unknown. The vertical scale represents the cumulative percentage of failed items or the probability of failure F(t) up to time t. The X-axis plotting position corresponds to the age at failure. The Y-axis plotting position probability of failure value is best estimated by the use of median ranks. Bernard's formula estimates the median rank of the cumulative probability of failures. The median rank estimated is preferred to the mean or average for nonsymmetrical distributions. The Bernard's formula for determining median ranks is expressed as

$$\text{Median rank} = (i - 0.3)/(N + 0.4) \tag{11.15}$$

where:
  i   is the failure order number
  N   is the sample size

The defining parameters of the Weibull line are the shape parameter β and the characteristic life or scale parameter η. β offers an idea of the physics of the failure that the item exhibits, such as infant mortality, random, or wear out. It equals the slope of the Weibull plot line on the Weibull plot. The scale parameter η, also called the *characteristic life*, equals the time for a

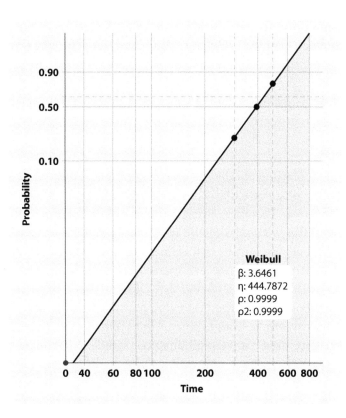

FIGURE 11.5
Weibull probability plot.

probability of failure of 63.2% for every value of β. Let's assume our caustic water pump suction strainer has experienced three failure events due to the failure cause "low suction pressure due to a dirty strainer," coded 2-A-2-a in Figure 7.5. If the ages at failure for the suction strainer are 505, 401, and 298 h of operation, we can determine their corresponding median ranks to construct the Weibull plot. The following order of events is recommended for performing Weibull analysis when precise age data is available:

1. Collect and use good data
2. Reorder failure data in ascending order
3. Determine median ranks for each failure event
4. Plot age to failure (X) versus median rank (Y) for every failure event
5. Draw a best fit line for the set of plot points
6. Determine the shape parameter β
7. Determine the characteristic life η

TABLE 11.3

Caustic Pump Strainer Failure Data

Failure Event Order (i)	Failure Time (h) (X)	Median Rank (%) (Y)
1	298	20.63
2	401	50
3	505	79.37

If the first step is completed, we may now reorder the events and determine the corresponding median rank value for every failure event, as shown in Table 11.1. We can construct a Weibull plot manually or through the use of software.

The data of Table 11.3 is then plotted in Weibull analysis software, as shown in Figure 11.5. The shape parameter β equals the slope of the Weibull plot line (3.65 in this case). Also, the scale parameter η is easily obtained by inspecting the graph for the time corresponding to a 63.2% probability of failure. A horizontal dotted line crosses the graph exactly at 63.2%, and its intercept with the plot line yields a characteristic life of approximately 445 h. If we substitute both parameters into the Weibull's reliability equation, our mathematical model for this particular failure is represented by the expression

$$R(t) = e^{-\left(\frac{t}{445}\right)^{3.65}}$$

Its corresponding probability of failure counterpart is expressed by

$$F(t) = 1 - e^{-\left(\frac{t}{445}\right)^{3.65}}$$

But, there is no need to use this mathematical model to calculate the reliability or probability of failure values. Unreliability values for any age are determined by inspection of the Weibull plot as was done for the β and η parameters. For example, F_{10}, or the time for which there is 10% probability of failure, corresponds to an operating time of approximately 238 h by inspection. Bear in mind that Weibull analyses are carried out for single failure causes subject to unique operating contexts. That is, we cannot analyze obstructed pump inlet strainers of pumps with different operating conditions in spite of their being physically similar. By the same token, we cannot analyze clogged strainers and worn impellers in the same plot, even if they belong to the same asset.

TABLE 11.4

The Meaning of the Value of β

β Value	Physics of Failure	Consequence Management Policy	Comments
β < 1	Premature	R, F	T, C not recommended.
β = 1	Random	F, R, C	Consider R for high λ. C could be considered.
1 < β < 3	Random + wear out	C	Perfect scenario for C.
β > 3	Strong wear out	T	T is preferred over C.

How do we interpret the Weibull plot? In Chapter 3, we explained the meaning of the shape parameter, β. Table 11.4 summarizes the relation of β values with the physics of the failure for each individual failure cause. It also validates the recommended consequence management policies for four possible scenarios of β values. In this case, β > 3, meaning that we have a strong wear-out case, for which T type tasks are recommended. In fact, Figure 10.11 shows that a time-based task is to be performed by the operator for cleaning the pump suction strainer. Thus, the analysis of quantitative data confirms that the consequence management policy chosen for this failure cause is technically supported by the Weibull analysis. In using Weibull, we've added the benefit of evidence-based decision-making to our RCM-R® analysis. The RCM-R® analysis team could change the consequence management policy previously recommended, if its life data analysis were to confirm a different physics of failure. For example, getting a β < 1 would warn us about premature failures likely to occur due to bad operating/maintenance practices or external causes such as improper design. Then, a redesign task dealing with the current state should be added to the recommendations listed in the RCM-R® report.

PERIODIC TASKS FREQUENCY

Condition monitoring, time-based, and detection tasks are performed periodically as part of the asset's PM program. The RCM-R® analysis team determined not only the task but also its frequency based on their analysis using simple mathematical criteria. We discussed in Chapter 3 how T task frequencies are calculated when failure and cost data is analyzed.

The optimum replacement or *preventive maintenance time* is calculated by the use of the following equation:

$$CPUT(t) = \frac{\text{Total expected replacement cost per cycle}}{\text{Expected cycle length}}$$

$$CPUT(t) = \frac{Cp \times R(t) + CuX(1 - R(t))}{\int_0^t R(t)dt} \qquad (11.16)$$

where:

Cp   is the preventive replacement cost
Cu   is the corrective replacement cost
R(t) is the probability that failure will not occur up to time (t)

The preventive maintenance task cost Cp is only $50.00, and the failure cost Cu is about $1050.00 as defined in the failure effects (see failure 2A2 in Figure 10.10 for our pump strainer clogging failure cause). When solving Equation 11.16 with Cp, Cu, and R(t) for many replacement time values ranging from 0 to approximately 800 h, we obtain an optimum (tp) of approximately 151 h. The cost of replacing the filter every 151 h is approximately $49.04 per replacement cycle, whereas the cost of failures is estimated as $20.23. The component reliability R(t) for 151 h of operation is approximately 98%, meaning that only two failures are expected in 100 replacement events. The results of the optimal replacement interval output produced by reliability analysis software would look like this:

Cycle length: 150.345626
Replacements per cycle: 1.000000
Planned replacements per cycle: 0.980731
Unplanned replacements per cycle: 0.019269
"Cost" of planned replacements per cycle: 49.036531
"Cost" of unplanned replacements per cycle: 20.232846
Mean time between replacements: 150.345626

Type C task frequency can also be determined mathematically. The standard Naval Air System Command NAVAIR 0-25-403 uses the following equation to determine condition monitoring task frequencies. This formula can be used for the optimal condition monitoring task frequency calculation for random failures with nonsafety consequences:

$$n = \frac{\ln\left(\dfrac{-\dfrac{MTBF}{PF}Ci}{(Cnpm - Cpf)\ln(1-S)}\right)}{\ln(1-S)}$$

where:

n   = number of inspections during the PF interval
PF   = interval from potential failure to functional failure
MTBF = mean time between failures
Ci   = cost of one inspection (CM) task
Cnpm = cost of not doing preventive maintenance
Cpf   = cost of correcting a potential failure
S   = Probability of detecting the failure with the proposed CM task, assuming that the potential failure exists

Let's assume that our scrubber system fan is assigned a vibration analysis task to detect potential bearing defects for avoiding a \$100,000 (Cnpm) downtime incident. The cost of the condition monitoring tasks (Ci) is estimated as \$100. Also, the PF as reported by the vibration technicians is presumed to be approximately 4000 operating hours. The maintenance planner estimates the costs of a planned repair (Cpf) as \$5000 and a probability of failure detection (S) by the maintenance technicians of 0.925, based on the historical ratio of bearing potential failures to total bearing failure events, including both potential and catastrophic failures. In other words, 25 out of 27 bearing failures as revealed in the computerized maintenance management system (CMMS) historical data were detected prior to failure. If the fan bearing's MTBF is 20,000, the NAVAIR formula yields n = 2.4, meaning that the frequency of the C task should be set to every 1671 h of operation, as determined by Equation 11.17.

$$\text{PM frequency} = \frac{PF}{n} \tag{11.17}$$

We all know that a PM job plan with the exact frequency (Fpt) requesting a task every 1671 h Mpt is unrealistic. Therefore, we maybe need to accommodate the vibration monitoring task, doing it every 1500 or even

1250 h, in our PM job plan for practical purposes. Note that we would increase frequency, not decrease it, if we needed to make these adjustments.

D task frequency was discussed extensively in Chapter 8. Now, we have covered mathematical formulae to support all type of tasks (C, T, and D) included in the PM work orders.

CHAPTER SUMMARY

RAM analysis is an essential tool of RCM-R®, enabling the analysis team to define an asset's reliability, maintainability, and resulting availability under their current operating context. Ai, Aa, and Ao are the three forms of availabilities an asset, system, or plant can exhibit. While Ai considers only failure events, Aa includes both preventive and corrective maintenance events. Ao considers all types of downtime for determining the item's operational availability. Hence, you will experience that Ao < Aa < Ai.

Weibull analysis involves the creation of statistical models from failure events data. Failure ages versus cumulative failure percentage are plotted on a special log-scale paper. The two defining parameters of the distribution are the shape parameter β and the scale parameter η (also known as the characteristic life). They are easily determined by inspecting the plotted line. Weibull analysis is very useful in determining the physics of single failure causes. β is the more important of the two parameters, as it enables us to determine a suitable type of consequence management policy for treating each failure cause. Optimal T, C, and D task intervals are determined by the use of engineering formulae. RCM-R® supports task frequency by analyzing failure events whenever failure data is available in the asset's current operating context.

REFERENCE

1. Dr. Robert Abernathy, The New Weibull Handbook, 5th edition, Reliability and Statistical Analysis for Predicting Life, Safety, Supportability, Risk, Cost and Warranty Claims, Abernathy, North Palm Beach, December 2006.

12

Implementing RCM-R®

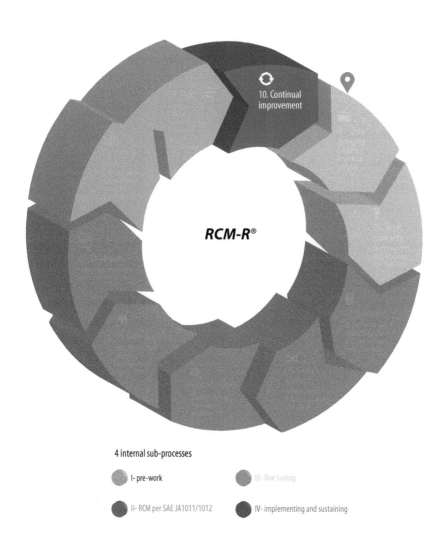

4 internal sub-processes

- I- pre-work
- II- RCM per SAE JA1011/1012
- III- fine tuning
- IV- implementing and sustaining

10. Continual improvement

RCM-R®

RCM-R® is a process that requires careful preparation, execution, implementation, and follow-up to ensure you get through the process, get the results you are looking for, and continue to see those benefits long after the analysis is complete. Experience shows that reliability centered maintenance (RCM) efforts, as with so many other initiatives, such as implementing new maintenance software, introducing lean management, six-sigma, total productive maintenance, or other "named" improvement initiatives, needs more than a purely technical team performing solid analysis to deliver results. Many think that change management is needed—it is, but there is much more required for success here.

This chapter focuses on those elements needed to make RCM-R® a long-lasting success in your organization. We present an approach that many years of trial and error have proved to work best.

THE ANALYSIS TEAM

RCM-R® is a team effort. It requires analysis performed by a team of people who are most familiar with the systems being analyzed. These team members normally carry out functions other than RCM-R® during their working days—operating, planning, supervising, engineering, maintaining, and so on. You may even have technical experts specialized in the system or equipment being analyzed, but none of these people are likely to come to the table with extensive RCM experience. Most will be new to RCM and specifically RCM-R®. They will require preparation.

Your team requires training and some practice with the method before they will be comfortable with it and competent at analysis work. Of course, you must also begin with the right people on your team.

The RCM-R® process draws on knowledge of what the system under analysis is expected to do (its functions), how it fails to provide those functions (functional failures), how it gets into those failed states (failure modes), and what happens when they occur (effects). It then requires analysis of consequences and decision-making about proactive and default tasks. If you have sufficient data that is fit for purpose, then it can also draw on your failure records (history).

Operators of your system, or operators of similar systems, are needed to answer the very first question in RCM-R® about functions. If the system is a new (greenfield) design, then process engineers who know how they intend the design to function will be needed. Likewise, we need that same expertise (operators or process/manufacturing engineers) when describing the effects of failures.

Maintainers and reliability engineers will contribute failure modes and have a great deal of understanding of proactive maintenance technologies and capabilities. If you end up with defaults such as redesign, then they are very helpful in developing ideas for how to deal with the specific situation—they are good "idea people." They are also the ones most familiar with your computerized maintenance management system/enterprise asset management (CMMS/EAM) systems and the data you have around failures, repairs, and your existing preventive maintenance (PM) program (if there is one). Of course, engineers also bring a degree of comfort with the mathematics used in manipulating that data when it is available to the team.

Specialists are available in some organizations, and they can be very helpful if the operators and maintainers get stuck on technical fine points. They may be specialized in condition-based maintenance (e.g., certified vibration or ultrasonic analysts) or specialized in a certain asset type or class (e.g., diesel engines or high-voltage transformers), and so on. As a general rule, you don't need technical specialists in your equipment and systems to have success with RCM-R® analysis. In fact, including them on the team full time is often resented, because they usually have a great deal of other work that demands that expertise. The RCM-R® analysis will only demand it when the rest of the team is stuck, and then, usually all they need is advice on a technical matter. Participation by the specialist is usually limited to answering questions outside of analysis meetings.

The most common successful team composition for RCM-R® analysis consists of one or two maintainers and one or two operators, full time, specialists (if available) on call part time, and a team facilitator.

THE FACILITATOR

Facilitators are your available experts in RCM-R®. They need not be your own staff, but if you are going to do a lot of RCM-R® analysis work, it is most economical to develop a few of your own. In many cases, companies opt to have facilitators from outside—usually from the same firm that delivers the RCM-R® training, and often, they use the actual instructor.

Facilitators do not need to be technical experts from maintenance or operators or specialists in the assets. In fact, it is usually advantageous if they know relatively little about the systems under analysis. That will keep them unbiased while they facilitate the analysis meetings. They should have a few key characteristics to be good facilitators:

- They must be very knowledgeable in RCM-R® (highly trained).
- They must be very good with the mathematical aspects of reliability analysis (or have someone very good with it on their analysis teams).
- They do not need to be engineers, or technologists, although most are. If they are not highly capable with the mathematics, then the team will need someone who is.
- They must have a technical background. There are many who argue that a good facilitator can be anyone, but we disagree. RCM-R® is looking at technical matters exclusively. The facilitator need not know the systems under analysis, but they do need to know technical principles well enough to understand the systems they are working on. It is not uncommon for analysis teams to get too wrapped up in technical details that will not truly impact the decisions that need to be made. The facilitator must be able to spot those circumstances so that she/he can move the team along.
- They must be very organized. The analysis itself can get complicated, and it is not unusual to have work assigned to various team members outside of the analysis meetings. The facilitator must keep track of all that.
- They must be very good "people people." They must be individuals whom your analysis teams will "get along" with easily. If they are people with prickly personalities, or they are highly judgmental, then they may struggle to keep the team focused.
- They must be familiar with and competent at using facilitation skills to run meetings and dealing with people and challenges that arise in a meeting setting (e.g., conflict resolution, breaking deadlocks, gaining consensus).
- They must be results oriented. While the facilitator is making sure a well-defined process is used, they must remain focused on getting good decisions from the team. The facilitator's competence plays a big role in the quality and pace of analysis that is achieved.
- They must want to be a facilitator. Volunteers only—draftees usually do the minimum they can get away with, and your RCM-R® efforts

will suffer as a result. Volunteers will also be people who like RCM-R® and possibly see career opportunities in taking on the role. However, watch out for those who only want it to add polish to their resumes or only to advance in their careers.

TRAINING AND COMPETENCY

Analysis team members need training in RCM-R®. We have seen people with no knowledge of RCM participate in analyses and get completely lost in the process. The thinking and logic used are not common knowledge, and they are not always intuitive, even for very sharp engineers. While the questions we ask are straightforward, the answers are not always so simple. The team members need to understand the questions they are being asked. All team members must also have the same understanding, or discussions will get very confusing. RCM-R® also has its own technical language. For instance, analysts need to understand C, T, R, 2, D, and F decisions without explanation. They need to understand the technical criteria for the C and T tasks. They need to understand failure finding concepts and what is "hidden" and "evident." That knowledge is best imparted in training. Even if an analyst has had training in some other form of RCM, they should take the RCM-R® training. Not all commercial variations on RCM use the same terminology or make the same use of analytical processes.

During the training, analysts are given case study exercises to perform. This familiarizes them with the worksheets we use and how the analysis is documented. When they join an analysis team, they are prepared, and the team can immediately begin analysis.

Untrained analysts will need to be taught as your team works through the analysis, step by step. The facilitator, who IS NOT a trained trainer, will need to explain concepts and the process as you work through it. That takes time away from the analysis itself and ultimately demands more time from the entire team. It is far more efficient to have all analysts arrive at the analysis meetings trained.

Facilitators are your specialists in RCM-R®. Their training is more involved and extensive—they need to show up at the analysis team meetings ready to lead the meetings confidently. There is no room for them to learn on the job, especially if they are working with a team of newly trained analysts. There will be plenty of opportunity for the analysts to make mistakes, and the facilitator will need to be confident and competent at catching them.

Some commercial variations of RCM provide long facilitator training courses. We have found these to be unnecessary and even a bit misleading. In these other variations, we see some flaws:

- The long course expends a great deal of time on case studies where facilitator candidates share the facilitation role. Usually, more than half the course time can be dedicated to this activity, and by the end of the course, none of them has actually facilitated a full analysis.
- Up to 25% of the course time is spent reviewing previously taught material from the analyst training. If the two courses are done in close proximity, this review can be overkill. After all, the facilitator candidate already has all the reference materials and previous training.
- New content of the courses is limited to discussion of facilitation skills and a more in-depth exposure to failure finding mathematics.
- The real learning happens in the facilitators' first analysis projects. Ideally, those are done with their trainer coaching them. If that is done, then the facilitator will emerge as competent. However, often times, they are not given this luxury. Many companies, having spent a great deal on a long training course, expect their facilitators to be ready to facilitate and fail to provide the coaching. That is a big mistake. Our experience shows that the quality of analysis from those who were not coached is lacking, often substantially. Those facilitators know they are doing substandard work but are usually powerless to get help. They lose interest and motivation and ultimately stop facilitating. This forces the company to train more—which in the long term is more expensive than provision of the coaching would have been.

For RCM-R®, we address those flaws head on. We use a training approach that relies heavily on mentoring and experience and not on classroom learning and shared facilitation of canned case studies.

Our facilitator competency process can be compared to the training of new pilots:

- Classroom instruction on theory and application.
- Practice flying, step by step (takeoff, landing, level flight, maneuvers, etc.) with an instructor sitting beside the student.

- Airborne testing: The instructor simply watches and corrects if necessary.
- Solo flight and final certification.

Facilitator candidates must be first trained as analysts and have performed at least one or two analyses before they move into the facilitation role. That weeds out those who don't really like it and don't want to do it as well as those who just don't quite fully understand the method. It also eliminates the need to review the earlier training (unless there is a long period of time between the original training and analyses and the beginning of their facilitation mentoring). In that case, they should redo the RCM-R® training. This gives them the theory and some practical application—like the first step of pilot training.

The facilitator candidates are mentored to develop competency—like the second step in pilot training. One-on-one mentoring by a very experienced facilitator/instructor ensures that the new facilitator gets the best possible learning experience. For their first analysis project, we "co-facilitate"—sharing the work. This allows the candidate to see how an experienced facilitator works, providing a good example and setting the standard. When the candidate feels confident enough to take over that role, he does so, and the mentor provides only advice and tips when needed. This swapping of roles occurs at each step in the RCM-R® process. By the end of his first "real life" analysis, the facilitator has had a chance to facilitate each step in the process, has probably made a few mistakes and been corrected, and has delivered a valid RCM-R® analysis result that can then be implemented in the organization.

The co-facilitating is then carried out for a second analysis. This time, the experienced facilitator (mentor) remains entirely in the supporting/mentoring role, while the candidate runs the entire analysis on his/her own. By the end of it, the candidate is usually very confident and has the necessary competence to facilitate on his/her own. This is the equivalent of the third step in pilot training—the mentor provides the candidate with a written assessment of his/her performance and any last tips before the facilitator goes solo.

The final step is the third analysis by the facilitator on his/her own. Our mentor remains available (usually by phone or email) to answer questions if the new facilitator gets stuck and needs help, but otherwise, he is allowed to fly solo. At the end of that third analysis, we feel confident that the new facilitator is competent enough to be certified.

PREPARATION

RCM-R® analysis begins before the team analysis meetings start. There is preparation work to be done, which is usually performed by the facilitator.

This preparation includes meeting room and scheduling logistics, availability of technical information, availability of trained team members, and preparation of operating context.

Team availability for analysis work can be a challenge. In most cases, you will be using maintainers and operators from an operational facility. This analysis work is being done in addition to their normal duties, unless the organization can find a replacement and dedicate the analysts to this work (something that rarely happens any more). Also, operators and maintainers in most operations do not control their own daily schedules. It will be necessary to set meeting times that the team can accommodate, so scheduling of the meetings can take a bit of effort and some negotiation with analysts' bosses and planners/schedulers. Note that there may be occasions during the analysis when the team runs into unknowns, and research is needed outside of analysis meetings. Normally, this does not require a lot of time, but team members and their bosses should be cautioned that there may be additional (minor) demands on their time outside of the meetings.

Because the analysis can be somewhat tedious, especially to people who are normally very active in the field (i.e., your maintainer and operator team members), we recommend that meetings be limited to 3 or 4 hours at a time with at least one or two short breaks. It's a good idea to schedule analysis meetings for half days and let your analysts go back to their normal jobs for the other half of the day. When feasible, *fast track analysis* pilot projects can be done, for which the team and the facilitator need to be available to work 8 hours per day during several days until the analysis is finished. The author has used this analysis mode widely, especially when in-house analysts are being trained to eventually become certified facilitators. Also, fast track projects are carried out when the analysis is needed promptly. For either case, the organization should make provisions for covering the fieldwork the RCM-R® analysis team constituents are leaving behind. Usually, fast track projects take 6 to 10 days of analysis, depending on the system's complexity, after which a complete report is compiled. Companies relying on external facilitators conveniently use the fast track mode, saving them a lot of money on unnecessary traveling time.

An experienced facilitator can actually manage two analyses in parallel—say, one in the morning and one in the afternoon—but he must be very good to do that successfully.

One challenge faced early in the process is to estimate how many meetings will be needed—see the next section on estimating for guidance.

The analysis meeting room should be relatively quiet and in a location free from disruption, but close enough to the system installation (if practical) to enable the team to visit the asset during the course of analysis work. It should have comfortable seating for team members (they'll be there for 3 to 4 hours at a time) with desk/table top space for them to review drawings, take notes, and so on. The room should be capable of being darkened and have a screen and a projector. The facilitator (in most cases) will project the information and decision worksheets as the analysis is progressing and he is recording answers to the RCM-R® questions.

In some cases, the facilitator may have a scribe assigned to do the recording or use flip charts. Whatever the facilitator chooses to use, he/she must organize it for the meeting room. If the meeting room is shared with other users, it will need to be booked—reserved at times when the team members can be available to meet.

There are a few analysis "tools" we find useful to facilitators. We have a handbook of tips and tricks for their use, copies of the decision diagrams, and even some short video clips that can be used as refreshers for the analysis team at each step in the process. Of course, the facilitator also needs a computer on which to record the analysis, and it must be equipped with appropriate software. We often use spreadsheets and MS Word documents, but there are a variety of RCM software packages available on the market that can be used. Caution is warranted with any software package—most of these are designed to work with specific commercial brands of RCM and not necessarily with RCM-R®. While the methods that comply with standards SAE JA1011 and JA1012 should all be very similar, there may be differences in how their associated software packages work.

When considering software, there is one key point to keep in mind. RCM-R® is an analytical process that depends on thought and decisions made by humans, not software and not computers. It is important to record the analysis in a logical way for future reference and use, but it is not particularly important that any one software package or another be used.

The analysis will refer to a variety of information sources. Ideally, these are all available at the fingertips of the team, but we have found that this isn't always necessary. If maintenance and reliability records are kept online in a CMMS or an EAM, then a terminal that provides access to that is useful. If those systems can generate ad hoc reports, then you will need someone who knows how to do that, either on the team or readily available to the team when needed.

Various documents such as P&IDs, schematic diagrams, wiring diagrams, instrument loop diagrams, mechanical drawings, civil drawings, technical manuals, procedures, standard operating procedures (SOPs), standard job plans, and so on may also be needed. Again, they should be available, but aside from the P&IDs and technical manuals, most can be accessed if and when needed. They don't all need to be in the room. If much of this information is stored electronically, then access to the document management system will be needed, along with someone who knows how to use it.

The operating context is part of the analysis itself, but as with any document, creating it with a team can be tedious and time consuming. One role of the facilitator is to prepare a draft of the operating context to the best of his/her ability before the meetings commence. It will be reviewed and amended as necessary by the team in the first analysis meeting. Keep in mind that the facilitator is probably not an expert on the system being analyzed, so there may be some research required, and possibly even interviews with a few individuals who are familiar with the system to enable preparation of the draft operating context.

ESTIMATING THE EFFORT

Experienced facilitators can have a quick walk around a system and review its P&IDs, and then come up with a reasonable estimate of how many analysis meetings will be needed. Inexperienced facilitators usually have no clue how long it might take. To begin with, they don't quite know how fast they'll be able to go, how well the team will answer questions, how much explanation the team may need, or even how deep the analysis will need to go. They may also look at many components in a system and imagine that each requires a great deal of detailed work. The experienced facilitator faces all those challenges with one big difference—he/she has done it before. After a few analyses have been done, estimating becomes much easier, and a very experienced facilitator can make it seem easy, almost an art.

The key is in one unknown variable and one variable that is only partially known.

The unknown variable is the number of failure modes you will encounter. As a rule, the more complex the system is, the more failure modes there will be. However, redundancy of equipment (duty and standby), multiple instances of the same equipment or item (e.g., heat exchangers or pumps), complex assemblies that are never disassembled and repaired, and so on all impact the "count" of failure modes. Indeed, seemingly very large systems can actually be fairly small in terms of the number of failure modes that will be encountered. During training we attempt to give teams a sense of how much effort is required for analyses, but it is during the mentoring phase that the facilitators will gain a great deal of "feel" for how to do this. Unfortunately, the only way to know with certainty how many failure modes you will encounter is to actually carry out the analysis and then count them up. That is why this part of the estimation effort is more art than science.

The other variable is partially known—the experience of the facilitator himself/herself and the team members. A new team with a new facilitator will move slowly. An experienced team with a new facilitator will move a bit faster. A new team with an experienced facilitator can move quite quickly, and an experienced team with an experienced facilitator will move very fast. Keep in mind that even experienced teams and facilitators can get "rusty" if they haven't done any of this work for a while. They may be slower today than they were 6 months ago when they last did an RCM analysis. The pace of analysis can vary from 2 to 12 failure modes per hour to do the entire analysis. A system with 120 failure modes might take 60 hours or as few as 10 hours to complete. The art is in knowing where on that scale you'll be.

Knowing the experience level and how recently your experienced teams and facilitators have last done some RCM-R® work is an important consideration when scheduling your analyses. If you intend to cover a number of systems, you will find it more efficient to do them in relatively short order—don't leave long gaps between the analyses, or you'll find the teams move a bit more slowly, getting less done than they might otherwise have accomplished in the same time frame.

Fortunately, the analysis has a couple of timing points where you can take stock of your progress and determine whether or not you will need more, or even less, time to complete the work. Again, during the training we offer to our facilitators, we share those precious gems of information.

Indeed, we use them in the mentoring process so that the facilitator gets a good sense of how to do this in real life.

CONDUCTING THE ANALYSIS

Once the preparation work is done, the draft operating context is available, the team arranged and trained, and the meeting room prepared and available, the analysis begins. The team will meet.

In the analysis meetings, the facilitator asks questions and records answers (or has a scribe do so). He can use flip charts or computer technology with projection so everyone can see what is being recorded. It is important that they do see what's being written down. The use of a computer with projection is best, in our opinion.

When recording answers, the facilitator is well advised to be careful with his wording and spelling. It is easy to be sloppy, thinking he will get back to it later and "clean it up," but that is often not so easily done. The facilitator also has another day job to perform! If she is diligent and careful in how she records the analysis, then very little work needs to be done outside of meetings. Of course, if flips charts are being used, progress will likely be slower, and there will be a lot of transcribing work to be done outside of analysis meetings—it works, but it is not recommended.

In the first meeting, you'll review and agree on operating context as well as "rules" for meeting conduct (e.g., no smoking, no cell phones, one speaker at a time, importance of sticking to agenda, use of a parking lot for after-meeting work, etc.).

In subsequent meetings, the facilitator will quickly review what happened in the last meeting, get the inputs from any "homework" and incorporate those into the analysis work, outline how far you hope to get in each meeting, and reinforce the rules. Analysis picks up from where you left off at the end of the previous meeting.

Following the analysis, the team can be dismissed to their normal duties, but the facilitator still has work to do. He will arrange for a review or audit of the team's work, arrange to present findings to management, and make sure the implementation of results occurs, as described in the bullets below.

The facilitator runs the meetings and may need to deal with any one or several of a number of potential challenges that can arise. Among these are:

- Keep people focused: End "side bar" discussions; you may need to remind people of their purpose there from time to time.
- Keep the team positive and motivated: The easiest way to do this is to stay positive and enthusiastic yourself—set the example.
- Watch the energy in the room, and take breaks if needed.
- Remind people of the "rules" if they disrupt the meeting.
- Stick to your schedule and objectives for each meeting: Enforce punctuality for start, finish, and breaks.
- Close out discussions if you are running out of time.
- Recognize when the team just "doesn't know" and needs to get help.
- Use parking lot to get information the team is clearly lacking (usually, this is evident, because discussion goes in circles).
- Shut down discussions arising from personal agendas (e.g., someone wants to steer the analysis to keep old practices in place).
- Resolve conflicts among team members (and stay out of them yourself).
- Help team get to consensus: Everyone doesn't have to agree, but everyone must be able to live with the decisions.
- Get quieter team members to participate (ask them directly for their thoughts or input).
- Tone down the inputs of more outgoing team members (can ask others if they agree with her/him).

These are all basic facilitator duties and skills. Some people are naturally good at this, and some are not. They are all skills that can be learned with a bit of instruction, mentoring, and practice. Another role that does take specific technical competence is to help team members who have forgotten part of their own training, misunderstand some of the RCM-R® concepts, or get confused (e.g., failure modes and effects often get confused; many people are confused about hidden vs. evident failures). This is why the facilitator needs all the training that his/her team members get, plenty of practice, and then the mentoring in his/her early days as a facilitator.

IMPLEMENTING THE OUTCOMES

Once the RCM-R® analyses are completed, then the resultant decisions need to be implemented. The analysis team is made up of those who most likely act on most of these decisions in the field, but they are not usually the ones who manage that work. That falls to managers, supervisors, and planners, who may not be participants in the analysis.

Someone must make sure that the decisions get put into practice, and that "someone" is usually the facilitator. A key facilitator role is to ensure that the follow-up on all RCM-R® decisions gets done in a timely manner. If he fails to do this, your RCM-R® efforts result in little more than binders or computer files full of analyses that achieve nothing at all.

The outputs of RCM-R® need to be converted into forms, standard jobs, plans, inspection routes, and so on and entered into whatever management systems are in use to ensure execution of those decisions as intended. Those in the field who will execute the work (e.g., perform the daily routine checks, keep equipment clean, monitor conditions, etc.) will need to understand and accept the changes being asked of them. It is not uncommon for resistance to arise, especially if an old PM program that was heavily reliant on T type tasks is transformed into a program with many more C and F tasks. In one electric utility, we realized the resistance came from a lack of understanding of the concepts behind T and C tasks. We offered training to those who were resisting, and then they understood well enough to begin accepting the decisions. Some even asked to participate in analysis sessions and became valuable contributors to the effort!

Others may find it difficult to accept that running some items to failure is now acceptable, especially if the past "culture" was one of firefighting reinforced with heartfelt thanks and appreciation for extraordinary repair efforts. The heroes won't feel the same if there are fewer fires to fight. Those who depended on all the chaos to earn more pay on overtime will be genuinely disappointed. Upfront efforts to manage expectations are invaluable in preparing people for what they will encounter as a result of RCM-R®.

We need to deal with these very natural human responses to change. Failure to do so puts the whole effort at risk, and experience shows that this is where most of these initiatives are most likely to fail.

Setting expectations, informing people in advance, and training them so they'll understand (and maybe contribute) all help. As the initiative progresses, keep people informed about progress, significant findings, and changes that will come as a result.

We find that you can never train too many people. For starters, the more people who are fully aware and informed, the fewer are likely to challenge the outcomes. Those who have had training but haven't participated in analyses can also make valuable contributions elsewhere—including systems that will not be analyzed (perhaps because they are of lower criticality).

The training results in new ways of thinking about equipment (i.e., what it does is more important than what it is) and what a failure really is (i.e., loss of function, not necessarily totally stopped) and how it fails (i.e., looking to causes for answers). Even without formal analyses, trained employees can, and do, see flaws in existing PM programs and operational practices. If they are encouraged to speak up, and management actually listens, then things can improve quite dramatically—both in the technical sense and even in employee attitudes toward their work.

The more you can get involved in the analyses, the better. Rotate people through analysis teams. If they participate actively, then they are part of the decision-making and are far more likely to want to see their decisions implemented.

RCM-R® analysis meetings are also great events for knowledge transfer and capture of corporate memory. Including younger workers in the team together with more experienced ones results in a great deal of learning. The experience is transferred in discussions, and much of it is captured in answers to the RCM-R® questions. In several places, we have encountered workforces with two distinct age groups—older baby boomers who were close to retirement and younger generation X and Y employees with less experience. The work ethics and motivation of these two cohorts are quite different, and many books and articles already deal with these. We have found that the gen X and Y employees learned a great deal from their baby boomer team mates and gained a healthy respect for their experience. Conversely, we also found the baby boomers impressed with their younger colleagues' quick thinking, ease, and comfort with using online tools to research for information the team needs and their generally higher level of education. The two groups usually find themselves getting closer and become more like one team than two separate groups. This improvement in teamwork and attitudes among workers is an often unexpected spin-off benefit from RCM-R®.

IMPLEMENTING C AND T TASKS

These tasks are maintenance activities—mostly performed by maintainers, but possibly some by operators or others. RCM-R® outputs a brief task description and a task frequency and specifies who is best to do it.

Following the analysis, you will have long lists of tasks for various pieces of equipment and systems. They are all destined to be done by the same group of maintainers and operators, all working within the same PM program.

Task consolidation is done to group tasks by trade, frequency, and geography. For example, you may have several system analyses that each generate several oil sampling requirements on a variety of equipment, at various frequencies, in the same plant, and all likely designated for the same trade (e.g., oiler). Rather than a standard PM work order being created for each one of them, they can be grouped. All the monthly samples are put on one PM, the quarterly ones go on another, and so on. This consolidation can be done system by system or plant by plant. The lists for those "sampling routes" can also be changed as more analyses are completed. It does not make sense to wait until all analyses are done before you start taking action on the decisions. Routes such as the oil sampling route described in this paragraph can be modified with each new analysis that is completed.

We achieve this consolidation by sorting the task outputs by type, trade, and frequency. Where it appears that consolidation makes sense, we do it. Where tasks are best left on their own, then we leave them on their own. This effort can be carried out by planners after the analyses are completed.

Another step that cannot be missed is to plan the work. Most C tasks are nonintrusive inspections, checks of readings, or sampling activities. There's unlikely to be much, if any, need for materials or parts, but you will need to specify the correct vibration analyzer, which infrared camera, the correct sample bottles to use, and so on. You also need to tell the maintainer (e.g., the vibration analyst) what level of signal is acceptable and what signal puts us in alarm. It is also necessary to specify what to do when the signal goes into alarm—for example, if vibration reading exceeds 0.35 ips, then initiate work order for bearing replacement.

T tasks normally intrude on the equipment and require it to be shut down. These repairs or restoration activities will usually consume parts and materials and require de-energizing of systems and operations to make the equipment ready to be removed from service. In some cases, they may also require operations to make alternative arrangements to continue production while the equipment is out of service for PM. These jobs require complete and comprehensive job plans. Again, it is the planners who take on this planning effort once the RCM-R® outputs are known and approved. Each of these plans should be saved as a "standard job" in your CMMS/EAM system and should be associated with the time- or usage-based PM trigger for the task execution.

C&T tasks will form the basis of your new PM program for the assets you've analyzed. Any old PMs should be modified to match the new criteria or deleted so that they are no longer performed. RCM-R® does not always "add to" an existing PM program—in fact, it can often remove tasks from it. In one electric utility that was faithfully following manufacturer's recommended maintenance programs, we found that we were eliminating as much as 30% of the work they had recommended and either running to failure or replacing the tasks with condition monitoring. The overall PM program actually got smaller than it had been and cost the utility 20% less per year to execute while increasing asset reliability! It is not uncommon to see PM programs being reduced by an even greater margin, ranging from 40% to 70% of time, when the original manufacturer-recommended PM programs were used. A recent case documented in a water treatment plant yielded a 63% reduction in PM man-hours and about 55% of reduced corrective maintenance cost after 1 year of the RCM-R® plan implementation. The plant found a lot of (unnecessary) maintenance-induced failures. Also, the ratio of T to C tasks was dramatically reduced as a result of the analysis.

IMPLEMENTING D TASKS

Detective maintenance (D) tasks (failure finding tests) are similar to C and T tasks in terms of analysis follow-up activity by the facilitator. Many of the D tasks are destined to be carried out by operators—after all, the best failure finding tests actually simulate the failure situation and see if the protective device actually does its job. The operators are in the best position to do this. They can keep an eye on process performance while they are simulating failed conditions. If the protective devices don't actually function as they should, then the operators can return operations to their normal state and get the protective devices repaired.

These tests are often logically grouped by frequency, but because of their potential to disrupt specific system operations, they should generally be consolidated system by system.

Planning for these tests is also important, but in the case of tests carried out by operators, instead of having a PM work order generated automatically, most organizations include the testing in SOPs. SOPs may be written and amended by process engineers, probably not by the planners.

The facilitator has a role to play in making sure that these testing requirements are defined to the right people so that they can be incorporated into the SOPs. Remember, too, that the new testing protocols replace old protocols (if there were any). In modifying SOPs, it is also necessary to delete older tests that may no longer be relevant or that may conflict with the ones defined using RCM-R®. The only caution we offer here is to be careful not to delete anything that is a result of some regulatory requirement without first getting a release from the regulators.

Regulators are primarily interested in safety and the environment. Whether or not your organization makes money is quite irrelevant to them. If your analysis work produces testing requirements that differ from the regulations, then you have a good basis on which to change those tests. You can always increase testing without getting permission. Doing something in excess of regulated requirements is never frowned on, but if you find your test is less frequent than a regulation calls for, then you'll need to get the regulator/inspector to grant you a waiver. Your RCM-R® analysis will give you good reason to make the changes, but they must still be convinced.

F OUTCOMES

Running to failure is a common output from RCM-R® analyses—as many as 30% or more of your failure modes may end up this way. As the term implies, doing nothing (proactive) is what these decisions mean to your organization. But you must still carry out repairs when they do fail. For each of these, you will need to have planners create a standard job plan that will be saved and ready for the eventual and inevitable failure that will occur. When used, those job plans will be included with routine (nonrush) work orders for the repairs.

But, there is a bit more to consider. When an F decision is made, it is because the analysis has revealed that the organization can in fact tolerate the failure. Anyone who participated in the analysis will know that. When an operator who participated in the analysis sees the failure occur, she'll create the maintenance work request with a standard (no rush) job priority. However, what about the person who wasn't involved in the analysis? In his case, he will also create a maintenance work request, but because he isn't aware that the asset can be allowed to fail, he might give the work request a high (rush or emergency) work priority. Maintenance will respond differently to these two work requests—one will be treated as routine, and the other might be treated as nonroutine. Unless someone in the maintenance planning group catches it, the rush work request may result in parts being rushed in at high cost, late-night calls to supervisory or management staff, time spent calling workers on a seniority list, taxi costs to get your workers into the plant, overtime work, and so on. All of this results in high costs for something that isn't really all that urgent.

Organizations need a way to tell their operators and planners that an item has been analyzed and it has been deemed acceptable to allow it to fail. It might be a simple list that is posted in operations control rooms so operators can assign the correct work priority, or a list that is available to all the planners so they can adjust work priorities accordingly, or a note/code in the CMMS/EAM that indicates the asset as subject to an F-type decision.

R DECISIONS

Redesign decisions are somewhat trickier to follow up, because they are for actions that need to happen only once, and they will usually have to be done outside the operations and maintenance departments. The facilitator, usually someone from operations or maintenance, is usually not a part of the department that must implement these decisions.

R decisions manifest in modifications to procedures (SOPs), practices (e.g., shop practices for housekeeping or separation of precision work from welding and cutting), training (often managed by HR or a separate training department), changes to standards (usually engineering), or actual changes to asset design (usually an engineering or even vendor responsibility).

The facilitator will have a list of R decisions coming from the analysis effort. Usually, these are defined in concise terms (e.g., "add oil sampling techniques to operator training" might be a result of a failure mode caused by oil contamination that was undetected using older/existing sampling methods). Having been a party to those decisions, the facilitator is in a good position to describe them more fully for the departments/individuals who will have to act on them after the analysis has been reviewed and approved. The facilitator will take those decisions, likely on separate lists, to each group that must implement them. The decisions are turned over for action, but the facilitator's role doesn't end there.

We cannot rely on others to act on something that we've "tossed over the fence" and into their department. They will need to know why they are getting these requests—part of the change management considerations we discussed earlier—and they will need to understand that they are not optional. Once the analysis is approved, all the decisions must be carried out, or the organization will continue to suffer the consequences those decisions were intended to deal with.

The facilitator will need to keep track of these decisions and who has them for action, and the status of the action item up to the point where it has been implemented. It is only then that we can be certain our R decision has actually resulted in the one-time action that is called for.

MONITORING AND CONTINUOUS IMPROVEMENT

Let's assume that the analysis was carried out successfully, all decisions were implemented, and results in terms of cost savings, improved reliability and productivity, and fewer accidents and environmental noncompliances are now expected. But, do we know we are actually getting them? Studies show that most improvement projects' results are never actually measured and that most don't actually deliver the results they were targeting. We want our RCM-R® initiative to be the start of a reliability culture in our operations, so it must be sustained. To sustain it, we will need to show without doubt that it worked and continues to work.

Let's assume you've been successful at starting that new culture in which reliability is valued over firefighting. Will it last?

Over time, things will change—operating context can change, and that can impact on functions, failure modes and their causes, as well as the effects of failures. These changes occur normally as a result of market changes in demand for product, customer behavior, changes in process inputs, modifications made in manufacturing processes, climate changes, growth in demand and load, and so on. These changes can occur suddenly (e.g., a process change in a manufacturing line), but more than likely, they occur slowly, over a long period of time. Consequently, we begin to experience a slow growth in unexpected failures and the need to react to them quickly. At first, we probably won't even notice this occurring because we are so busy, but in time, we will eventually realize that we are no longer getting the benefits of RCM-R® that we once realized.

Failure modes we thought we were preventing or predicting are cropping up more than we expected. Our C tasks are not catching as many incipient failures early enough. Our T tasks are being done, but failures are increasing anyway. Our D tasks are still finding failed protective devices, but a few hidden failures have been missed, and the multiple failure occurred. More infant mortality failures are occurring. It appears that the RCM-R® efforts are no longer working.

One organization found that after about 6 years from completing its analyses, failures were increasing, and in one case, they were sued by a customer as a result of a single dramatic failure that they had thought they were preventing. Over time, things changed—work execution discipline had slacked off and PM completion had fallen, loads on some assets in the field had increased dramatically due to growth in customer demand, and the effects of aging had been understated in the initial analysis. It took a lawsuit to trigger a hard look at these factors:

- PM completion discipline was within their control by making supervisory staff aware of the important of PMs: A bit of training was also performed so that supervisors would understand why the PMs existed in the first instance.
- Load on the assets would only continue to grow: No one had thought that existing assets might need to be upgraded in time, but in fact, that is exactly what was needed. In this case, it was electrical loading, and older equipment simply wasn't rated for the growing demands.
- The increased loading on the older assets had actually accelerated the deterioration of those assets. The desired performance was relentlessly creeping closer and closer to built-in capacity, and in a few cases had exceeded it.

Catching these things happening as they occur, and not after a lawsuit is threatened, requires diligence in the long term. There is a need to monitor conditions that would have been underlying assumptions in analyses. These are generally stated in the operating context. In the example above, it was the operating context that was changing. As those factors change, then there is a need to revisit the analyses and make sure the decisions (which should be actioned in the field) are still appropriate. A monitoring and continual improvement program is needed.

Our RCM-R®-derived proactive maintenance program will include considerable condition monitoring (C tasks). Yet, the entire proactive maintenance program also needs condition monitoring to ensure it is still performing. Not long after we carry out our analysis work, we need to analyze our operating contexts and expected performance outcomes for assets and systems. From that, we need to determine what we must monitor in the long term to ensure the program overall is still delivering on its expectations.

Again, this task falls to the facilitator as your in-house expert. You may have a reliability department or engineer who monitors this sort of performance, but it is likely they are not monitoring the whole program overall. Doing that is one of the good asset management practices that are outlined in the new International Standard for Asset Management—ISO 55001.

MONITORING AND IMPROVEMENT TOOLS

In addition to the dedicated, watchful eyes of your facilitator(s), there are two primary tools for monitoring your programs. One is root cause analysis (RCA); the other is PM optimization.

Root cause analysis is used after we experience unexpected failures— either major failures that happen once or chronic failures that happen repeatedly. In root cause analysis, we observe the effects of a failure—in fact, it's the consequences of the failures that usually trigger RCA. We gather data about the failure itself, what happened, what sequence, who was involved, actions taken, actions that may have been missed, what actually broke, what was its failure mechanism, and so on. With that information, we look for possible causes of the failure, and we keep tracing back till we find a "root cause" that we can control. If we change that, then theoretically, we will eliminate the whole chain of events that followed it, including the failure itself. Doing this avoids those effects and consequences that triggered us to carry out the analysis.

RCA is a powerful tool, and there are a variety of different ways to carry it out, but they make use of the technical knowledge and decision-making logic we also use in RCM-R®. In RCA, we don't start by looking at asset functions as we do in RCM-R®. Rather, we begin with the effects of a failure, work back to identify the failure modes and causes, and then apply a logical decision process to determine actions that can be taken to eliminate those causes. In some variations of RCA, we actually use an RCM decision logic to arrive at decisions once we've identified the causes.

As with RCM-R®, the outcomes of an RCA analysis must be implemented in the field and monitored to be sure that it is working as intended.

RCA is a reactive approach to continual improvement. It will catch failure modes and causes that may have been missed in RCM-R®, but it only does so after the fact. It's highly successful, but sadly, that success depends on a failure to identify the failure modes and causes sooner.

RCA is not suitable for use in developing an entirely new PM program. You probably wouldn't choose to fly on a new airplane if you knew they were developing the maintenance program entirely by using RCA.

PM optimization (PMO) is another tool we can use, and it is more proactive than RCA. In PMO, we are periodically reviewing our PM program for flaws. As with RCA, there are a variety of different approaches to doing PMO. Like RCA, it also has a different starting point for analysis than RCM-R®. We don't begin by looking at functions; we begin with existing PMs that you may be using today. These can be derived in any way—from manufacturers' recommendations or based purely on in-plant experience or even a PM program that was developed using RCM or RCM-R®.

PMO lies somewhere along the spectrum between proactive (RCM and RCM-R®) and reactive (RCA) approaches to improving a maintenance program. It is often applied in "brownfield" (existing) operations where a PM program is already in place, but it can be used in a "greenfield" (new) application if you apply it to a proposed PM program (e.g., one that may be based entirely on manufacturers' recommendations). Unlike with RCA, we don't need to wait until we have failures to apply it. In that sense, it is proactive. It falls short of the level of proactivity that RCM and RCM-R® achieve, because it starts with an existing outcome rather than deriving it from functions.

In PMO, we begin with existing PMs. We look at what that PM actually can achieve and ask whether it is being used appropriately where it is currently specified.

For example, in a small power plant, we found a program of scheduled vibration analysis checks at 6 monthly intervals on several diesel generator sets. The vibration analysis equipment in use was an older displacement-type meter that measured mils of displacement. Diesel engines tend to vibrate—they are reciprocating machines with some inherent mechanical imbalances and counterweights and flywheels designed to smooth out their operation. Despite that, they still vibrate much more than most equipment with purely rotary motions (e.g., pumps and centrifugal fans). The displacement readings were useful for their pumps and fans, an oil centrifuge, and a few other pieces of rotating equipment, but not for the diesels. The readings they got were typically very high and really didn't indicate much that was relevant except that diesel engines vibrate more than other pieces of equipment (something they already knew). In fact, a vibration analyst will probably tell you that the large vibration inherent in reciprocating equipment renders it unsuitable for most forms of vibration analysis. The vibrations from piston movement and the many gears and chains have so much energy and "noise" that they effectively drown out smaller vibrations from the purely rotary components. We concluded that the displacement readings were not really adding any value, so that task was eliminated.

The other problem with the vibration program they had was the frequency of performing checks. They had a contractor come in to take readings at a 6 monthly interval. Vibration analysis can detect flaws that are still very slight long before they will become problematic. But, it requires sophisticated analysis methods to do that. The displacement readings they were taking would detect flaws and give them warning of incipient failures, but not as early as, say, acceleration readings might have done. The flaws they would find would be much more advanced, more severe, and closer to the point of functional or even total failure. The 6 monthly interval was deemed to be far too long to be useful in catching all failures. Indeed, they had experienced failures that were "missed" as well as successful catches. We increased monitoring frequency to monthly.

In that example, we optimized a vibration analysis program within some fairly common constraints—they couldn't have the contractor on site all the time, and they were limited by the contractor's technology. We eliminated some very low-value checks and increased frequency on those that were already proved to be valuable.

We did this by looking at what the PM check was actually capable of finding, questioning whether or not that was reasonably likely to be present in the equipment, and then deciding on whether or not to keep the task, and if we kept it, how often we should do it.

We also questioned the particular monitoring technology and its suitability for the application. If we had had the option of changing that, we might have made different decisions.

We didn't work back to identify all the failure modes in the equipment, but we could have made an effort to do that, too. Some PMO methods do make an effort to identify the more prevalent failure modes and then work forward (as in RCM-R®) to determine appropriate failure management policies using a decision diagram. In our example of the diesel generator sets, we worked back to identify potential bearing failures and machine imbalances as failure modes that could be detected using vibration analysis techniques. Had we not been limited by the available technology, we might have looked for other methods to detect those same problems (e.g., acceleration readings with full spectrum analysis to isolate specific component signals).

However, even if we had done that, we might well have missed other failure modes that were never addressed initially. For example, there were installed thermocouples on the cylinder exhaust manifold to indicate high exhaust temperatures (and hence valve problems) in specific cylinders. The thermocouples were part of a monitoring system that the operators relied on. However, there was no check of the monitoring system itself, except to depress a test push button that only really demonstrated whether or not the alarm indicators were functioning. It did not test the thermocouple and its electrical circuit, but the operators thought that it did. When we first looked at that test routine (it was tested daily), the operators thought it was a valid test and sufficient to detect problems, and on the surface it did appear that way. We insisted on digging deeper into the design of the monitoring system and its test button. In doing that, we found that, indeed, the actual alarm circuits were never fully tested. We devised a testing protocol for the thermocouples and resolved the oversight, but had we only looked at what the test push button was testing, we might not have thought of that.

PMO is a good way to check on what you are doing or proposing to do, but it is not as thorough as RCM-R®, and it can miss key failure modes if its practitioners are not diligent. It is a good way to start on a reliability improvement program in a brownfield application. First, optimize what you have in place so you stop wasting effort, and then go back over your critical applications using RCM-R®. Also, if you have a PM program that was developed using RCM-R®, then it is a very good way to check on your program periodically to see whether it is achieving its intended purposes.

PMO should be able to identify where changes in operating context now render past decisions somewhat invalid.

GOVERNANCE FOR SUSTAINABILITY

RCM-R® is an important step toward having a proactive "reliability culture"—the sort of environment you hear about at "world-class" performers and those who experience superior results consistently. It fits well within a program that includes RCA and PMO.

Executing RCM-R®, successfully implementing its outcomes into day-to-day operating practices, and even monitoring and sustaining its success over time (using RCA and PMO) require more than the technical and operational activities those entail.

Changes in personnel, especially at management level, can make quite an impact on the culture and what does and does not happen in any organization. In one large mine, a new maintenance manager was hired, who had a long career of success in operations where people were used to doing what they were told. In this new role, he had a better-educated and more empowered workforce to contend with, but he didn't change his approach. The mine had already begun an RCM initiative and was seeing some early success, but there was a lot of analysis work yet to be done. The new manager had read about RCM and had learned that it had a historically poor track record where people didn't follow up on decisions that were made. He had accurate information, but he expected his workforce to be like previous ones that had not taken initiative nor acted on results from monitoring programs. He assumed that his new workforce would be the same and that ultimately, the RCM efforts would fail. He cancelled the initiative! The program wasn't sufficiently advanced to produce enough evidence to convince the manager of its value (that was poor timing), nor did it enjoy sufficient senior-level sponsorship to survive in the face of his personal preferences (that was poor governance).

Programs like RCM-R® need to be in place for the long haul. It is not a project that you start, work through, complete, and forget.

Businesses are far from static—they experience frequent changes in management and suffer personnel turnover at all levels. Newcomers may or may not be exposed to RCM-R®. They may come from environments where break-then-fix was the norm, they may have very "old school" thinking about preventive maintenance, or they may focus on costs only. For the reliability culture to thrive and survive, it must be supported, as any long-standing program is supported, by corporate governance processes insisting it survive. In fact, if we look at what sort of programs consistently survive in business cultures, we'll see that they are actually nurtured so that they thrive. What thrives, survives.

Accountants have their routines and processes, they meet stringent guidelines and deadlines, and their work is subjected to audit. There are regulations and laws that demand these processes and rigor. Corporate decision-making follows rigid processes and practices at the board level to ensure the company meets regulatory and fiduciary obligations. Those governance processes sustain those practices in accounting and at the board level, regardless of who comes and goes and regardless of their preferences. Businesses sometimes consider the costs of doing this as "compliance" costs. Where there are regulations, that's pretty accurate, but more importantly, those practices that are being sustained are thriving because of the attention they get from regulators. That is because they add significant value—to shareholders and the public.

Safety and environmental stewardship have an array of regulations to follow. They tend to be prescriptive and aimed at avoiding the effects and consequences of failures. They rarely address the causes of failures. For example, you must wear personal protective equipment to protect against a variety of workplace hazards, but there is no regulation specifying that you must eliminate those hazards due to equipment failures. There seems to be an underlying assumption that failures are inevitable, and we must live with them. Yet, we engineers know better. We know that if we have a reliable operation, we have a safer and more environmentally compliant operation as well. Our focus on reliability addresses the causes of the failures that result in safety hazards and environmental noncompliances. Sadly, regulators haven't yet made the leap from addressing effects and consequences to addressing the causes, and as engineers, we have tended to move away from regulation as opposed to adding more. Consequently, that gap in thinking persists, but we don't need regulations to change it.

By our analysis work, we are starting on the path of developing a reliability culture. If we are to sustain its survival, we need to ensure it can thrive. What thrives, survives, so that means we must invest in it on an ongoing basis.

As in accounting and corporate board governance, our investment in the governance of our reliability culture must ensure that it survives turnover among individual key players while ensuring it is flexible and adaptable to changes in the business environment and company strategic goals and objectives.

Reliability governance is just as important as board governance and financial governance. Arguably, because of its potential to impact on safety and the environment as well as company financial performance, reliability is somewhat more important. Airlines, in response to an intolerable level of crashes in the early days of commercial air transport, responded with deep analysis and the creation of RCM. The subsequent reduction in crashes that occurred concurrently with the growth in air travel has no doubt saved many lives, making air travel the safest practical way to cover long distances (it's about 1000 times safer than road travel), and assured the long-term viability of a commercial airline industry. Arguably, the growth of air transport wouldn't have occurred had crash rates remained as high as they once were. The entire industry owes its existence to more than just flight technology—it owes it to reliable flight technology. In that industry, reliability is a part of the culture, and a good deal of governance is in place to ensure it survives. There are standards that must be applied. RCM (in its airline variation) must be used on aircraft systems, or the aircraft won't be licensed as airworthy. Armies of engineers and managers could change, but those practices will remain.

If we want a reliability culture, we need reliability governance. We need a corporate reliability charter defining how we will achieve that. It must have executive support and sponsorship. If the COO or CEO sponsors a program of reliability, then a new manager at a single site couldn't make a sweeping change that would degrade it without considerably more than his personal preference as his decision-making criteria.

How that governance will look in any given organization will be unique to that organization. Referring back to Chapter 1, reliability is at the heart of delivering good asset management. Our focus is on performance and balancing that with risks, costs, and opportunities. Reliability takes us there. With the introduction of international standards in asset management in 2014, we now have a framework for that overall asset management program. Reliability governance, as we've called it here, doesn't need to be called that. If good asset management practices are in place, then the necessary governance mechanisms will also be in place to ensure that our reliability culture thrives and survives.

13

Leveraging RCM-R®

Reliability centered maintenance (RCM) is a process that produces decisions, each of which must lead to action in one form or another:

1. Predictive maintenance (C) tasks
2. Preventive maintenance (T) tasks
3. Detective maintenance (D) tasks
4. Redesigns of the asset (R)
5. Changes to training (R)
6. Changes to procedures (R)
7. Changes to standard practices (R)
8. Do nothing (F or run to failure)

Each of the C, T, and D tasks will have a prescribed frequency at which the task should be executed. The last chapter dealt with implementing those tasks. The R outputs require one-time action only. Finally, the F outputs don't really require anything to happen, but it is wise to make sure that people know where F is considered the right choice—if you don't, then people might treat them as emergencies when in fact they are not. Again, Chapter 12 provides tips on how to deal with them.

In addition to implementing the outputs of RCM-R®, there is even more we can do to get the maximum value from our analysis efforts. In this chapter, we will deal with ways to leverage your RCM-R® outputs.

CONCEPT

The military has an integrated and iterative process for developing materiel and support strategy known as integrated logistics support (ILS). It aims to optimize functional support for systems, leverage existing resources, and guides systems engineering efforts toward lower life cycle costs for the installed systems and a minimal logistics footprint, making the systems easier to support. Today, it has migrated from purely military applications, where the authors first encountered it, into commercial product support and customer service organizations.

The concepts of ILS are behind your ability to get your car serviced without having to wait for parts and behind the mechanics' ability to diagnose faults in your car's electronics. Someone had to think about what faults could be present and what signals they would produce, develop a diagnostic tool to find them and present them to a technician, and train the technician on how to use the system and then on how to fix it once a problem is identified. Similarly, when you send something back to your supplier for repair, there is an entire support organization in place to service your need. When an airline sends a jet engine back to a repair facility for maintenance, they've got exactly what is needed and in the right quantities; they've got the trained technicians and all the equipment they need to do their work. Providing all of that so that it works relatively smoothly and on demand just when you need it is the result of a great deal of thought and preparatory action. Why not put that sort of efficient and effective support in place for industrial systems that you maintain and support yourself?

INTEGRATED AND ITERATIVE?

ILS is described as an integrated and iterative process. It is integrated because it considers all aspects of support, including parts, tools, test equipment, facilities, training, skills, information requirements, and so on. Identifying these requirements and putting them into place so they provide effective and efficient support requires recognition that each element can impact the others. For instance, the need for training in skills that might be new to the workforce that will have to use them could lead to the need for a new simulator in a training facility, and that, in turn, could lead to the need for maintenance of the simulator itself, a manual for it, parts for it, and so on. Every maintenance action recommended for a new system will almost certainly lead to the need to have a maintenance plan, parts, and tools. The need for parts will drive decisions about sparing and warehousing.

ILS is iterative, because it is first applied at the design stage in the life cycle of a new system. The analytical process begins with conceptual design. If flaws or support challenges are revealed during the early analyses of those designs, then they are addressed at the next, detailed design stage. The analysis is then repeated, potentially revealing even more challenges, typically at more granular levels of detail. The analysis isn't just looking at whether or not the system can function, but also at what it will take to support it in sustaining its functions.

For example, the author worked on a ship-building project using ILS at the design stage, before steel was cut. As the design developed, an early form of RCM was used to identify maintenance requirements. One system being designed into the ship was an electronic machinery control system. At the time, this was a relatively new concept for shipboard use (much like early distributed control systems in processing plants). The ship's crews had experience and training in older pneumatic machinery control systems but had no exposure to modern electronics. Because control signals were all processed inside one big "black box" instead of relays and pneumatic circuits installed in various locations depending on the system, troubleshooting would be different. In some respects, it could be easier using computers, but it would become far less intuitive to crews that were used to mechanical systems. It was evident, early on in the conceptual design, that a simulator would be needed for training purposes. That, in turn, would have to be housed somewhere, and the existing training facilities had no space. That led to a decision to build a new training facility to accommodate the simulator (and other new systems as well). As the design became more detailed, and the nature of failures that could be expected was revealed, the simulator design also evolved. Eventually, a manual had to be written for the new simulator (as well as the new control system on the ship), maintenance programs had to be developed for both, spare parts had to be identified and procured, and so on.

We were working on what was, at the time, leading edge technology. We had to develop it as we went along, making it an iterative process. Fortunately, many industrial systems are not so complex, nor are they leading edge, technologically speaking. They may already have supporting systems developed and working that we can take advantage of, but if you are a supplier of those systems, you will go through an experience much like the author did to develop these.

WHY BOTHER?

We are not suggesting that the complexity and high degree of thoroughness that the military applies to its billion-dollar systems is needed in every case, but some rigor and forethought are warranted. Without them, we are plagued with problems throughout the operational life of the assets that we rely on for our livelihood.

Consider the newly built plant. The project team has diligently collected technical manuals from equipment suppliers. They've kept track of purchase information, recommended spare parts list, maintenance instructions (if separate from the manuals), installation drawings, and other information. They've probably filed it all away in a project engineering library, and probably filed it by contract number, not necessarily process or asset number. They've bought the manufacturer's recommended commissioning spares and stored them in a laydown area somewhere near the new plant. Perhaps they've put them into containers or even a warehouse, and they've got a clerk looking after them. There is a drawing library of engineering drawings including schematics, P&IDs, installation drawings, plan views, elevations, as-built drawings in some cases, civil drawings, electrical drawings, and so on. These may be in hard copy and/or electronic format suitable for reading using dedicated engineering computer assisted design (CAD) software and printing on large format printers.

For design and construction purposes, this all works. But not so for maintenance and operations.

To begin with, maintenance and operations will have no insight into the contract numbers that vendors had assigned to their purchases. Finding information filed that way will be a challenge. Commissioning parts are a great way for suppliers to sell parts to those who are not giving their future needs much thought, but some of these will run out quickly if the supply chain hasn't been properly set up (e.g., parts catalogued and designated as stock or nonstock, min/max and economic order quantities determined, etc.). The technical manuals will contain generic operating and maintenance instructions. The operating instructions for a piece of equipment on its own (as described in its manual) will be different once that equipment has to be started up and run in a specific sequence during start-up of an entire plant. Chances are that the manual or instructions for start-up of the plant haven't been provided by anyone and may not even have been written. Likewise, the new plant is often lacking in procedures for shutting it down and dealing with emergencies, such as failures to key equipment that may occur. Special tools, lifting apparatus, and other provisions for maintenance are often overlooked—at least until those first few failures occur, and the need for them becomes painfully evident.

The new plant is in trouble even before it is started, and the troubles show up early. Like any new system, it will suffer from infant mortality problems as we start it up, in its early operational hours and days. We had better hope the commissioning spares cover all those instances.

What about the systems requiring maintenance that were not bought from a single vendor? What about systems that we built or that our contractor built for us? Do these have maintenance instructions, operating manuals, spares, or any documentation other than construction drawings? Likely not. If our organization depends on vendor manuals and instructions, we will be in trouble eventually, because those portions of our plant will be totally ignored. For instance, structural elements may be allowed to deteriorate until someone eventually notices how unsafe they've become.

Of course, the new operation will eventually stabilize. It will have gone through a lengthy start-up with plenty of teething problems and lessons learned. Each of these will result in some action to put in place the right documents, parts, tools, and so on—whatever was missed initially.

There are usually a lot of problems that can be traced back to poor hand-off of information, a lack of operator and maintainer input to design, and a focus on the project and its immediate outcome for as low a cost as possible. Little is spent on putting the needed support in place to make sure that the project delivers a truly operational asset that can be sustained in service. All too often, engineering has done its job, the contractors have done their job, and it is handed over to operations and maintenance to make it work. In doing this, a lot of modifications get made, and before long, the plant isn't what was designed or what was intended, but it works. Operators and maintainers deserve a lot of credit for making their facilities work as well as they do; it is often the result of a lot of hard work and lessons learned the hard way. And it could be so much easier!

INDUSTRIAL LIFE CYCLE SUPPORT

Let's now apply these concepts to our industrial systems. Figure 13.1 illustrates the process.

We begin the process of leveraging by applying RCM-R® to our system. Ideally, this is done at the design stage, because that is when you have the greatest opportunity to influence life cycle costs. Figure 13.2 illustrates a timeline to show when the ability to influence spending, and the actual spending are timed. They are not concurrent.

Costs are committed early in the design stages. Decisions made about design details can have a big influence on system operability and maintainability. Just ask any maintainer about how well the systems he maintains were designed. They are usually quick to point out flaws in design that lead to a lot of extra work to gain access, troubleshoot, and change parts. Designers are focused on providing a system design that meets design specs. Those specs deal heavily with operational output requirements and often ignore operability and maintenance needs. In fact, ergonomics in design is often treated as a specialized discipline. Historically, we've built mock-ups of systems and tested them to make sure they can be operated and maintained. Today, we have 3-D computerized modeling software, and we can simulate operational and maintenance activities using those models. If we use them, we can see where we need to place lifting lugs for slings, where we need to provide more overhead room to raise equipment off its mountings, where we need to provide platforms for maintainers and operators to stand, and so on. We can simulate difficult maintenance tasks to ensure they can be done without major disruption of adjacent systems that would cause excessive downtime for repairs. This is all within our capability if we choose to use it.

The time to do that sort of work is early in the design phases of the system's life cycle. Once we've entered the operational and maintenance phase of the life cycle (usually the longest phase), we can still make changes if we have to, but it is usually costly and often not done until we've already experienced the failure that leads us to discovering what could have been designed better. The later in the life of the asset, the less opportunity we have to influence remaining life cycle costs, and the payback or return may become too low to justify the costs. At that point, we simply "live with" whatever we've got and struggle along with whatever flaws it may have until we eventually replace it. For maintainers in this situation, we find ourselves hopeful that our engineers who design the replacement will

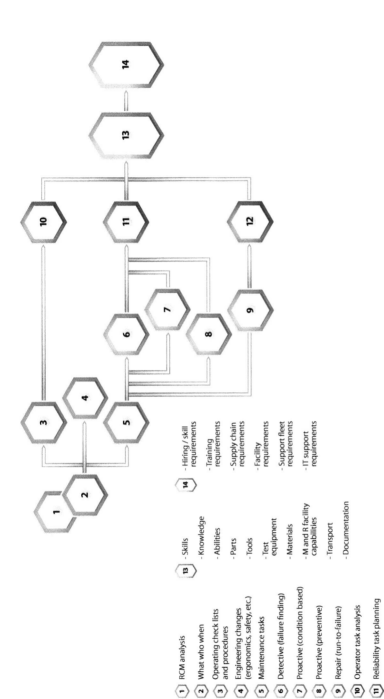

FIGURE 13.1
Industrial life cycle support definition.

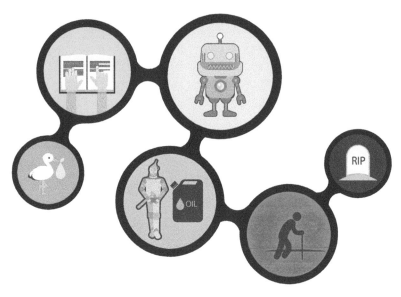

FIGURE 13.2
Opportunity to influence life cycle cost of ownership of a system decreases with age.

know about these flaws and actually do something about them rather than repeat the past mistakes. Sadly, that doesn't happen in all cases. There needs to be a feedback loop—something that good asset management practices (as discussed in Chapter 1) put in place.

Going back to Figure 13.1, we'll now look at the process steps, assuming that we are doing this early in the life cycle of the asset (preferably at the design stage).

The asset design is beginning to take form. As it does, we can apply RCM-R® to the systems we are designing. We'll get a sense of failure modes and what we can or might do about them. At this stage, we probably don't have a lot of detail about the design, just concepts and line diagrams—a pipe from this tank to that pump but no detail on the pump design itself. We don't need that detail at this stage. As we look at systems holistically, we'll see how they can fail and make decisions about whether we can live with those failures. We might see, for instance, that valve placement at particular points in the piping enables (or disables) desired operational functionality. We'll get a sense of how big the systems are going to be. For instance, a 1000 gpm pump is going to require headroom with some substantial lifting provisions and space for moving it around as part of its future maintenance. Our later, more detailed design will need to provide for this. This is all done by analyzing system designs using RCM-R®. We don't even attempt to analyze equipment at this point, because we don't know for sure what equipment we will be installing; we only know its primary and some secondary functions—but that is enough for our needs at this point.

As we move into more detailed design, we can update our systems' analyses. We'll have more detail on functions and definition of some of the specific hardware at this point. We'll be more able to identify failure modes at a more detailed level. Again, we may identify actions that will be needed that could influence design R outcomes from our analyses. We will also be getting C, T, D, F, and other R outcomes that require procedures and actions by operators and maintainers.

As we define the outputs from RCM-R® and gain confidence that our design is stable, we can begin to put the supporting logistics into place for maintaining and operating our system once it is put into service, beginning with maintenance and operator tasks. Each task has a prescribed frequency. If the task requires parts, then we can determine how often we will need those parts, and that gives us a basis on which to build our spares inventory and replacement strategies for when we eventually begin to use those parts. But, how do we know we need parts?

Each task defines an action. That action is planned in detail, listing parts, tools, technical drawings, work permit, and other requirements essential to executing the task successfully. So, for each task we define (C, T, D, and F), we develop a plan.

CONDITION MONITORING SUPPORT

Often, our C tasks will not require parts, but they do require the use of condition monitoring equipment—say, a vibration analyzer or infrared camera. These items of support equipment go on a list of needed equipment together with a frequency of usage. For example, if a vibration analysis is called for on a particular bearing once per month, then that is added to the usage of that same vibration analyzer for all other bearings. If we have a lot of bearing analyses to do, based on how many we are monitoring and how often we monitor them, we can determine whether we need one or maybe two vibration analyzers. We can also see clearly that someone needs to know how to use the vibration analyzer. If our required vibration analysis requires a great deal of knowledge for interpreting the signals that are read by our analyzer, then we can also identify the level of training the analyst will need.

TIME-BASED TASK SUPPORT

T tasks usually require some form of intervention. Even oil changes require consumption of oil and possibly filters at the same time. We can evaluate how much oil we need per change, how many filters, and how often we need them and use that information to define how much oil we need to carry and how many of which filters we will need on hand. Likewise for any scheduled overhauls—parts and other consumables can be defined with precision, along with their demand rate.

Tools and other support equipment can also be defined, just as they were for the C tasks.

If our T task requires that equipment be replaced when it is removed for overhaul work (e.g., a mining haul truck wheel motor overhaul), then we can also gauge how often we will need the spare and how many we will need. For the haul truck wheel motors, we can figure out how many we should carry in our warehouse to support the planned component change-outs across our entire fleet of trucks. Let's say we have 10 trucks with four wheel motors each that all require change-out every 18,000 h. If we operate our mine around the clock, those trucks might run for 20 h a day, reaching 18,000 h after 900 days (or a little less than every 3 years). So, every 3 years we require 40 wheel motor replacements. If we can safely assume that we'll be able to space those out over the 3 year period, we'll need spare motors four at time, about every 3–4 months. If overhaul time for those is less than 3 months, then the overhauled motors should return to us in time for the next needed replacement, but some may be found to be beyond repair. Replacement of these might take 6 months, so we would be wise to carry some spare motors in addition to a set of four that will cycle among the trucks. Just how many we carry in our spares inventory is covered later.

DETECTIVE MAINTENANCE SUPPORT

Detective maintenance involves testing of backup and safety systems, normally dormant, to see that they are still working. Unless the test method requires particular test equipment or other support, we might find that we have very little support requirement for these tests. But we will need support for the repairs that will result from finding the systems inoperable.

If we need support equipment for the test, then we will need it at the frequency of the testing. If we don't, then we only need to plan for the repairs to systems that we find failed. Of course, we will also need a forecast of just how often we will find them failed—that is, another RCM-R® result. In fact, we use the tolerable level of risk (i.e., how often we can tolerate finding them failed) in the calculation of the testing frequency. Again, our planning will tell us what we need, and our analysis will tell us how often.

RUN TO FAILURE (F TASKS)

F tasks require us to carry out repairs when the asset is allowed to fail. In a typical RCM-R® analysis, you can expect about 30% of your decisions to be F tasks. For support requirements, our repair plan includes identification of parts, and so on, and in what quantities per repair. The failure frequency (which we should also know from our RCM-R® work) tells us the demand rate on each part for each repair. This, too, feeds the spares calculation.

PROVIDING SUPPORT

A job plan for each RCM-R® output task should identify needed

- Parts
- Tools
- Test equipment
- Skills
- Diagrams, drawings, or other information
- Transportation needs
- Facility capabilities (e.g., access and lifting provisions)

If the job requires that other equipment or systems be de-energized or down, then these will also be identified, leading to support work by operators in making the systems ready for maintenance work.

Once these requirements are all identified, they can be provided as a part of the overall support package for our system.

Parts requirements will be defined by job, including identification of which part, how many are needed for the task to be carried out once, and how often we can expect the task to be needed. Often, we find the same parts used on multiple pieces of equipment in our facility. The frequency for demand on such a part is the sum of demands from all instances of its occurrence. If we don't consider this, we could easily overstock on parts. We handle materials and other consumables needed for the tasks in a similar manner. Lubricants, adhesives, gasket material, shims, and so on are effectively consumed in the conduct of maintenance tasks. They must be supplied for use and replenished as used.

Tools are handled like parts, except that they are not normally "consumed" in the repair task. They are reused for many repairs. Common tools are normally dealt with as property of the technicians who use them, but some special tools that are critical to the work but used only occasionally may be kept and managed in a "tool crib." Lifting equipment, chain falls, lifting straps, and so on are usually managed in a tool crib, because they must also be maintained and pass test requirements to be used safely. Again, we can estimate the usage of those special tools and determine whether we might actually need more than one on hand. For instance, a tool crib supporting a shop doing a lot of engine repairs may require more than one torque wrench. Perhaps one wrench per vehicle bay or for every two bays may be needed. If we do have tool crib items, then we also need a tool crib (if you don't already have one).

Test equipment is really just another tool, but often with highly specialized purposes. Today, it is often computerized; it may be static or mobile (e.g., on a cart). It may be sufficient to have one tester per shop, or if it's a large shop, to have several.

Special tools and test equipment also require maintenance. They may require periodic calibration to ensure their accuracy. In some businesses, calibration may be essential—for example, measuring tools used in pharmaceutical or biomedical applications. Tools used to test parts and systems will likely need to be calibrated. Calibration is sometimes regulated, especially in food, pharmaceutical, and biomedical applications. The use of tools requiring calibration can lead to the need for calibration equipment and possibly even a calibration shop. The calibration equipment itself also needs maintenance and calibration, often using a set of standards. For example, weigh scales are calibrated using a set of standard weights that must be kept under specific conditions for use in calibration. Storage and maintenance of these conditions while stored is another requirement of the support system that must be put in place and managed.

DOCUMENTATION AND RECORDS

In addition to having all the right things and in the right quantities when needed to execute the work that is called for, we also need some documentation. This includes drawings, diagrams, schematics, technical repair manuals, operating manuals, written procedures, check lists, and so on. Even the parts lists and plans we generate from our analyses are part of the documentation that is required. The documents we use must be fit for purpose—accurate representations, so that we can rely on them. They need to be up to date with any changes that may have been made.

As we execute work, we may need to keep certain records. For instance, we may need to keep accurate records of all calibration work in a pharmaceutical plant or all proactive maintenance that is executed in a nuclear plant. We will want records of readings taken during condition monitoring work so that we can monitor trends in what we are reading over time. We will want a record of any design changes or modification we make to equipment and consequently to parts lists for the support of that equipment we've modified. Modifications also cascade into changes in our maintenance plans.

All of this information is useful if it is accurate and current—fit for purpose. If it is inaccurate or incomplete, then we can be misled and make incorrect decisions, use the wrong parts, and cause failures that could potentially be harmful to production, service delivery, safety, the environment, or the reputation of our organization.

Consider that for all our physical assets and systems, we have a corresponding set of information (documents and records). That is our "information plant." Not only does this information require a "home," but it too must be maintained. How to do this is beyond the scope of this book, but recognize that our efforts will generate content for the information, and like our physical assets, it too must be kept up to date and accurate, fit for purpose.

We need to consider on- and off-line storage of information in a variety of formats and forms. We need to consider that the information will be used by many people in different places at different times and for different purposes. There is a need for only "one version of the truth" when it comes to this information, and that version needs to be the most current and accurate it can be. It also needs to be accessible, or it is at risk of being ignored. In the absence of accurate and valid information, we humans tend to "make it up" or go with whatever rumor we heard or whatever we can remember from the last time we had to deal with the topic. All of that is highly likely to be invalid in some way. The right information needs to be in the right place where needed and easily accessible to whoever will need to use it.

Consider, for example, the true story of a pulp and paper plant that operates a fleet of pumps. The maintenance plan calls for a C task—monthly vibration monitoring—used to make decisions relating to bearing health. When the vibration readings reach a critical level according to the vibration specialists, the bearing is changed out. The company wants to do even better. They want to develop an optimal policy balancing cost and risk based on the local operating environment. They won't be able to accomplish this sophisticated goal without having collected and stored the correct data. In this case, they needed to have kept records on previous bearing changes and their reasons, as well as vibration readings specific to each bearing location, including the date, time, and measurements. They were lucky that they had kept all these records and it was possible to improve the maintenance policy. Many companies are not so lucky, and despite collecting and storing what they believe to be large amounts of data, they are often missing key data elements, making it impossible to implement better data-driven maintenance practices.

Asset information management is a big topic requiring more coverage than this book can provide. It is often done poorly and has often led to troubles, big and small, because of what someone didn't know. The purpose of information is to help us make good decisions. We need to provide ourselves with good information to achieve that. The types of decisions we will make throughout the life of our assets dictate the information we will need. That, in turn, should reveal where information is needed, by whom and in what form or format. In turn, all this drives the storage and retrieval systems we use, including our information technology. If our mobile field technicians need accurate technical manuals in the field, then they need either a small library in their trucks that will be hard to keep up to date with modifications or some mobile technology for retrieval and reading of the information they'll need to use.

SKILLS AND CAPABILITIES

Work plans define what actions are needed in the conduct of tasks we defined in our RCM-R® analysis. These actions may require skills and knowledge that our work crews already have, or possibly new skills and knowledge. As part of our leveraging activity, we want to make sure our workforce are ready for their work—they may need training in the new skills and the knowledge they'll need to use the new condition monitoring technologies we've recommended.

For example, if we are going to use infrared thermography, vibration analysis, oil analysis, and/or ultrasonic technologies, we may need to train our maintainers or technicians on how to use them. Depending on how extensively we intend to use those technologies, we may need extensive training. If we intend to use vibration analysis as both a condition monitoring and a diagnostic tool to aid in finding exactly what problem we are detecting, then we may need more training than if we are using it only to detect that there is a problem.

RCM-R® will produce tasks for operators to perform. Some of those tasks are simple maintenance tasks, but our operators may not have any knowledge of how to carry them out correctly without training. For example, we may ask operators to perform oil level checks and to top up oil sumps when needed. They will need to be taught how to check levels, what levels are acceptable and what are not, what oils to use in which pieces of equipment, how much to use, and even how to spot abnormal conditions such as emulsified oil in a sight glass that indicates the presence of water.

For every plan we create, and every operator procedure or task, we should evaluate the skills that will be needed against the skills the work crews already have. If they need additional skills or knowledge to do what we are asking them to do now, then we need to provide training.

FACILITIES

Facilities are there to support the work we do. In fact, they often support the support work itself! Maintenance shops provide the infrastructure and systems we need to carry out maintenance—a support effort. Do our maintenance facilities have the capability and capacity to accommodate the new work we will be demanding to have performed there? We need to consider this.

In one northern mining operation, a fleet of haul trucks was replaced with newer and larger-capacity trucks. The larger hauling capacity meant that fewer trucks were needed in the replacement fleet, so they could avoid a bit of capital spending on trucks. However, the trucks were larger than the doorways leading to the maintenance bays in the truck shop. In the summer months, this was an inconvenience and somewhat of a joke about short-sightedness, but once winter came around, it became a major problem. At that mine, temperatures could drop to $-40°C$ ($-40°F$). If the trucks were stopped at those temperatures, the engine blocks would freeze. Working outside on precision work is not an option. Mine production suffered for several months while emergency modifications were made to the shops to allow the trucks to get indoors. It took several additional months to make other modifications to air, oil, greasing, and drainage arrangements and to acquire the right tools for work to be carried out efficiently.

Similarly, if we identify training requirements, we may need to consider our training capabilities and facilities. Do we have the needed facilities, simulators, or instructional apparatus to carry out the training? If not, we need to arrange for these as well.

Storage space for spares is another consideration. We know we'll need to carry some stores, so the need for a store room is fairly obvious. But how big does it need to be? Do we have spares that may need maintenance themselves? Some electronic spares (and there are more and more of these all the time) need to be powered up, even in storage. Some items need calibration before being issued for use. Do we have the ability to identify those and make sure they are calibrated? Rotating equipment spares (e.g., electric motors) need to be turned periodically. Do we have a system to manage the maintenance of these items while they are in storage?

SPARES

Neil Montgomery

Spare parts, lack of them, or having the wrong ones are often blamed by maintainers for their inability to get work done. Investigation into this often reveals that the maintenance planning was inadequate, and supply chain managers didn't have good information about what spares, how many, and when needed. We are hoping, through this process of leveraging our RCM-R® analysis fully, to avoid all that. There are a number of considerations to keep in mind with spare parts, some of them quite technical. The part can't just fit into the space where an old one was removed; it must also do the same things and to the same standard. We need to figure out how many we need, whether or not we can get them shipped to us in time to meet demand, and if not, then how many to keep and order each time, how to store them, what (if anything) we need to do to maintain them while in storage, what to do with them if we modify equipment and change the part in some way, and so on.

When determining the correct number of spare parts to keep available, the strategy first depends on how many spares are consumed, and how quickly. This is often related to their reliability and criticality to the whole process.

Many spares are disposable items that are rapidly consumed, easily available, and relatively inexpensive to purchase and store. There may be significant overhead costs involved in managing the ordering of such parts. For these cases, it would be most common to implement an economic order quantity or min/max stock level model. The accuracy of these models depends on the high demand rate.

Some parts are not consumed very quickly, but can nevertheless be so critical to the operation that access to spares is essential, either to be stored on site or by arrangement with a local vendor. The erratic demands for these spares mean that the calculations used for fast-moving parts may not be accurate. The remainder of this section will deal with the problem of stocking these capital or emergency spare parts.

Demands for a capital spare will come from two main sources: T tasks and F outputs. T task demands, briefly addressed earlier in the haul truck wheel motor example, happen when components need to be removed on a planned basis for some period of time for repair, for refurbishment, or possibly to be discarded. F output demands happen at the unexpected failures that may still occur from time to time despite the best efforts of a maintenance program.

It may also be possible that the demand for a spare part might be predicted through condition monitoring. This is a more advanced spare provisioning strategy that is becoming more popular, but we will not address it in this section.

Since the demand for spares due to T tasks tends to be predictable (and more frequent), the procurement of these spares should be considered as a separate management problem from the problem of keeping sufficient stock for much less predictable failure demands. In the haul truck wheel motor example, there were a total of 40 wheel motors that each needed a 3 month overhaul every 3 years. In an established operation, we can expect to spread this work evenly over the year, requiring about four spare motors just to allow us to operate this planned maintenance program in a steady state.

Note that in a completely new operation, once any infant mortality problems have worked themselves out, we can expect a period of very low demand for capital spares for both planned maintenance demands and unplanned failure demands. How long this low demand period will last depends on the reliability of the part and the number in service. Consider the wheel motor example again. What if they were all brand new? These items tend to be reliable. Perhaps only eight of them will fail before the first planned overhaul. It could very easily happen that there are no demands for spares at all in the first year, followed by a few failure demands in Years 2 and 3. Then, we will have to plan for 32 overhauls to come due after 3 years, all at essentially the same time. It is particularly important to manage T task spare demands separately from F output spare demands in the case of a completely new operation.

How many spares should be kept available for the unpredictable failure demands? Again, there are many possible considerations. Whenever one asks "what is the correct number?" one has to specify "according to what criteria?" Different criteria can give different answers. The company must decide which criteria best match its business objectives.

One criterion could be what we call "interval reliability," which is the chance we will never be short of a spare when it is needed over some specified planning horizon. We would typically want this probability to be high, such as over 90% or 95%. The planning horizon might be the lead time to source a new part, or it could be a fiscal year, or a 5 year plan, or whatever the business feels is suitable. Interval reliability depends on the demand rate for spares (essentially, the failure rate for an individual part times the number of parts in use) and the repair rate, if the parts are repairable.

A common error for companies using the interval reliability criterion is to maintain a stock equal to the expected number of demands over the planning horizon, but this tends to result in an interval reliability of only between 65% and 80%! The correct number tends to be the expected number of demands plus a buffer for the worst-case scenario.

Another criterion could be to stock the number of spares that minimizes total costs. Costs could include any of the following considerations: the cost of acquiring the part, holding costs, downtime cost per unit time if a spare is not available, and the cost of obtaining an emergency part at short notice. One should also consider the residual value of an unused spare at the end of a planning horizon.

Yet another criterion could be to stock the number of spares that supports a certain (presumably high) percentage of uptime. This criterion also depends on the demand rate and the repair rate.

The actual calculation for the number of spares requires the use of software. There are software packages available that can perform spares calculations. Care must be taken to understand exactly which calculation methods are being used by the software you select. Some companies use simulation software for spares calculations.

Let's revisit the haul truck wheel motor example. We will use the spares management software developed by the Centre for Maintenance Optimization and Reliability Engineering (C-MORE) at the University of Toronto for our calculations.

There are 40 wheel motors in use. The mean time to repair is 3 months. After 3 years of operation, we accumulate 1440 months of wheel motor use, during which time there were eight failures serious enough to require a spare motor to be used. This gives us a MTBF of 180 months per wheel motor. To achieve an interval reliability of 95% for a 12 month planning horizon, we would need three spares.

Suppose, instead, we wish to minimize the total cost per month including downtime costs and holding costs. We estimate a loss of $1,500,000 per month if a truck cannot operate due to the lack of a wheel motor. The holding cost for a spare is $2000 per month. The number of spares required in this case is four.

Or, suppose we wish to have all trucks operating 99% of the available time. This requires only one spare. Of course, the trucks can and will suffer downtime for reasons unrelated to the unavailability of a spare wheel motor. This calculation only concerns the contribution of the lack of spares. People are often surprised at the low number of spares needed to assure uptime. This is usually because critical items tend to be highly reliable, so that the repair time is only a small fraction of the MTBF.

Note that in each of these example calculations, the number of spares required is always over and above the four spares we need on hand just to service the time-based overhauls, which we have advised should be managed separately.

This expanded wheel motor example was possible because of the accumulation of data over 3 years of operations, which a new operation will not have. We have already noted some of the spare parts challenges faced in a new operation because the first operating period is so different from the steady state situation. In addition, one must add the challenge of no available data. The business may have to depend on original equipment manufacturer recommendations.

CHAPTER SUMMARY

We've learned about RCM-R® and how its enhancements can directly add to RCM analyses, linking them to risk management, concentrating efforts where they will do the most good, and making them more accurate (and more relevant to specific failure modes) through the application of Weibull analysis. From our military project experience, we've learned about a very common sense approach to leveraging upfront design analysis, and RCM results contribute to the setting up of a complete support infrastructure that is very specific to the system and failure modes being managed. We've adapted integrated logistics support (ILS) concepts to the broader industrial marketplace. Our RCM-R® analysis is iterated through successively more detailed design stages to reveal potential design flaws and corrections, where provision may be needed for access and support equipment, and what major insurance spares should be considered. Even in the early design stages we can reveal the need for support systems such as simulators, trainers, and even facilities. As design progresses in complexity, our depth of analysis can increase. resulting in a very precise definition of support requirements right down to spare parts and their quantities. Doing this helps us achieve what we typically see in military applications and new aircraft: relatively trouble-free commissioning and start up, rapid ramp up in capacity, and quick return to service if start up teething problems are encountered. Our systems come up to full utilization sooner, resulting in a more rapid payback on their capital investment. The best time to do this is at the design stage, when we have the greatest influence over life cycle costs and hence over long-term profitability of the new system. A simplified ILS process, industrial life cycle support definition, is provided. Using this helps attain a great part of the value that is intended to be achieved through the new international asset management standards, ISO 55001. Support for all RCM-R® outputs can be defined, acquired, and positioned where it will do to the most good. This includes spare parts, materials, tools, test equipment, training (new skills where needed), training support (e.g., trainers and their accommodation), lifting and transport capability/capacity, shop capacity and capability, technical documentation to support repair as well as training, record keeping systems, CBM equipment, and all the training needed to utilize support equipment and tooling correctly. Doing this extended analysis, leveraging on the very specific outputs of RCM-R® analysis, sets your organization up for long-term, successful, and sustainable operations, achieving exactly the system performance that you were hoping to achieve. It turns that hope into a real choice.

Appendix A: Failure Finding Task Frequencies

Failure finding tasks are used to find hidden failures. Hidden failures occur in protective devices providing some sort of protective or backup function.

In normal operation, the protected function is operational, and the protective device is dormant in a standby mode and available to operate if and when called on. If the protected function fails, the protective device should operate and serve to reduce or eliminate the consequences of the failure.

However, if the protective device has already failed at the time when the protected function fails (i.e., it is unavailable), then a multiple failure is said to have occurred. The combined effect of the two failures is that the protected function is lost, and the consequences that the protective device was intended to avoid will occur.

Because protective devices are normally dormant, their failures are seldom evident to operators—they are hidden. Once they fail, they will remain that way (unavailable) till detected. That will happen when the protected function fails and the protective device is called on to operate (and doesn't) or when it is tested to confirm that it is still working.

The probability of a multiple failure occurring is the product of the probability of the protected function failing (giving rise to a demand for the protection) and the probability that the protective device is unavailable.

The mean time between failures of the protective device (MTBF) of the protected function is the demand rate, and it is often known from maintenance records, operational history, or the history of similar systems in other applications. We will consider the average time between demands to be the MTBF of demand, or *Md*.

If we have not been testing our protective devices, we may have no idea how unavailable they are, but we might have a sense of how often they can fail (or have failed historically). Again, their MTBF (*Mv*) may be known, or we can find it with a bit of research.

Note: If failure is either age or usage related, then it can be dealt with using TbM, and we can ignore failure finding. However, if the failures are random (as they usually are), then we need to determine testing intervals for the protective device to reduce its unavailability to a level we can tolerate. If the failures are random (i.e., failure pattern E), then they have a constant failure (hazard) rate λ. This rate is the inverse of the MTBF.

For devices that fail randomly, their reliability (i.e., probability of survival) at any point in time, assuming it has survived to that time, is given by

$$R(t)=e^{-\lambda t}$$

Its unavailability at that point in time is

$$u(t)=1-R(t)=1-e^{-\lambda t}$$

The term $e^{-\lambda t}$ can be expanded into a power series:

$$\left(1+(-\lambda t)+\frac{(-\lambda t)^2}{2!}+\frac{(-\lambda t)^3}{2!}+...\right)$$

If we disregard the higher-order terms and substitute into $u(t)$, we are left with the approximation

$$u(t)=1-\left(1+(-\lambda t)\right)$$

$$u(t)=\lambda t$$

Substituting MTBF for λ, we get

$$u(t)=\frac{t}{M}$$

SINGLE PROTECTIVE DEVICE: RISK-BASED DECISION CRITERIA

Substituting *Mv* for *M*, the instantaneous unavailability of our protective device becomes

$$uv(t) = \frac{t}{Mv}$$

We are interested in its average unavailability over a period of time between testing intervals, I.

To determine the average, we integrate the equation over a period of time from 0 to I and divide by the length of the interval I:

$$Uv(I) = \int_0^I \frac{uv(t)}{I} dt = \int_0^I \frac{t}{I \times Mv} = \frac{I}{2Mv}$$

which is simplified as

$$Uv = \frac{I}{2Mv}$$

Solving for I, we get

$$I = 2\,Mv\,Uv \qquad (A.1)$$

This is the general formula for determining the testing interval of a single protective device that fails randomly.

The probability of the two failures occurring at the same time $(1/Mm)$ can be calculated using the product of the probability of the protected function failing and the probability that the protective device has failed:

$$\frac{1}{Mm} = \frac{1}{Mv} \times \frac{1}{Md}$$

Rearranging,

$$1/Mv = Md/Mm$$

Substituting into Equation A.1, we get

$$I = 2\frac{Md \times Mv}{Mm} \qquad (A.2)$$

SINGLE PROTECTIVE DEVICE: RISK-BASED CRITERIA AND PROBABILITY OF FAILING ON TEST

If the probability that the test could fail the device is p, then we need to consider two failure events and two components of unavailability. For the failure of the device not due to the testing, unavailability is the total unavailability minus the unavailability caused by the test:

$$Uv = \frac{Md}{Mm} - p$$

The probability of the protective device itself failing for reasons other than testing (Po) is lowered by p to become

$$Po = Pv(1-p)$$

Expressed as MTBF,

$$\frac{1}{Mo} = \frac{1}{Mv(1-p)}$$

Mv then becomes

$$Mv = \frac{Mo}{1-p}$$

Substituting Uv and Mv into Equation A.1, we get

$$I = 2\frac{Mo}{1-p}\left(\frac{Md}{Mm} - p\right) \tag{A.3}$$

Note that as $p \longrightarrow 0$, then $Mo \longrightarrow Mv$, and we are left with

$$I = 2 \times \frac{Md \times Mv}{Mm}$$

SINGLE PROTECTIVE DEVICE: ECONOMIC-BASED DECISION CRITERIA

This applies where we have operational or nonoperational consequences. We need to determine the relative costs of carrying out the failure finding test versus the costs of the multiple failure. Start with the basic formula Equation A.1:

$$I = 2 \times \frac{Md \times Mv}{Mm}$$

The probability of the multiple failure is the inverse of its MTBF per unit of time:

$$\frac{1}{Mm} = \frac{I}{2 \times Md \times Mv}$$

Its cost per occurrence is *Cm* (including the cost of repair and downtime), so its cost per unit of time is

$$\frac{Cm}{Mm} = \frac{Cm \times I}{2 \times Md \times Mv}$$

The cost of doing each failure finding task is *Cff* each time it is done. The cost of failure finding per unit of time is

$$Cf = \frac{Cff}{I}$$

When you find a failed protective device, you repair it at a cost of *Cv/Mm* per unit time.

Adding the cost elements, we get total cost

$$C = \frac{Cm \times I}{2 \times Md \times Mv} + \frac{Cff \times Mm}{I} + \frac{Cv}{Mm}$$

We want to minimize total cost *C* at time *I*, so we take the derivative and set it to zero:

$$\frac{dC}{dI} = 0 = \frac{Cm}{2 \times Md \times Mv} - \frac{Cff}{I^2}$$

Rearranging and solving for *I*,

$$I^2 = \frac{2 \times Mv \times Md \times Cff}{Cm}$$

$$I = \left(\frac{2\,Mv\,Md\,Cff}{Cm} \right)^{1/2} \tag{A.4}$$

MORE COMPLEX CONFIGURATIONS

If there is only a single protective device or a simple protective system that can be treated as a single device (which is often the case), then formulae A.1 through A.4 are all you need.

However, your protective device may be made up of multiple components having different failure rates (MTBFs).

It is important to consider the "logical" connection of devices more than the physical configuration. If devices that are physically connected in parallel with each other must both operate for a protective system to perform its function, then they are logically in series from a reliability perspective. Devices only operate in parallel from the logical perspective if they are truly redundant to each other. When determining failure rates (or MTBFs) for series and parallel combinations, the mathematics is somewhat different.

In reliability for devices that fail randomly, we are dealing with exponential functions. In series combinations, the failure rates of the components add together to comprise the failure rate of the combination. Mathematically speaking, we are adding up the exponents. Keep in mind that failure rate is the inverse of MTBF (which is more commonly measured). When adding a series of failure rates expressed as MTBF, we can't just add them together.

Series of Devices Making Up a Single Protective Device (System)

If the protective device is actually a series of devices that are connected in series (e.g., a sensor, wiring, a battery, and a light or alarm) or it has multiple failure modes, then its failure rate is the sum of all the various devices' (failure modes') failure rates:

$$\lambda v = \lambda 1 + \lambda 2 + \ldots$$

Expressed as MTBF,

$$1/Mv = 1/M1 + 1/M2 + \ldots$$

$$Mv = 1/\left(1/M1 + 1/M2 + \ldots\right)$$

Remember that $Uv = Md/Mm$, so the failure finding interval, $I = 2 \times Uv \times Mv$, becomes

$$I = 2 \frac{Md}{Mm} \times \frac{1}{\frac{1}{M1} + \frac{1}{M2} + \ldots}$$

$$I = 2 \frac{Md}{Mm} \frac{1}{\sum_{i=1}^{n} \frac{1}{Mi}} \qquad \text{(A.5)}$$

where:
   n is the number of series devices or failure modes
   Mi is the MTBF of each device

For the cost-based formula, substitute $\dfrac{1}{\frac{1}{M1} + \frac{1}{M2} + \ldots}$ for Mv

$$I = \left(\frac{2 \times Md \times Cff}{\sum_{i=1}^{n} \frac{1}{Mi} \times Cm} \right)^{1/2} \qquad \text{(A.6)}$$

Parallel Combinations of Redundant Protective Devices

Parallel combinations of devices are truly redundant. Unlike a series combination, in which any device failing takes down the entire system, when connected in parallel, both devices must fail for the system to lose function. Adding redundancy to any system design is a powerful design tool for increasing reliability, although it comes at a cost—you must add more devices (capital cost), and you must test them all (operating cost). When the consequences of failures are potentially severe and intolerable, the costs are justifiable due to the risk mitigation being provided by the redundant devices. However, merely adding redundancy without justification can be a costly design approach. Using highly reliable devices also works well if such devices can be acquired and used in your system design. However, keep in mind that a large number of devices connected in series, even if each of the devices is highly reliable, can become an unreliable combination. Parallel redundancy has its place, and testing them to make sure they haven't yet failed is a failure management approach.

There are two possibilities for testing redundant devices. They can be tested separately from each other or at the same time. If they are tested separately from each other at an average interval of I, then the average availability is slightly higher than if they were tested at the same I together. If they are tested together, which is in fact more likely, then I must be a bit shorter to achieve the same availability.

First, we look at testing them independently of each other (the less likely scenario).

Two Parallel Redundant Devices Tested Independently of Each Other

Each protective device added in parallel can fail, leaving a period where it is unavailable. The unavailability of one device is Uv, so the unavailability of two providing protection is now Uv^2.

The probability of multiple failure now becomes

$$Pm = Pd \times Uv^2$$

$$Uv = \left(Pm/Pd\right)^{1/2}$$

Since $P \propto 1/M$, and using Mt as a target value for $1/Pm$,

$$Uv = \left(Md/Mt\right)^{1/2}$$

$$I = 2 \times \left(Md/Mt\right)^{1/2} \times Mv$$

The constant 2 arises because failures are random and assumed to fall halfway between tests on average. To achieve this, testing intervals must average to I. For this, the independent parallel devices should be tested independently of each other and not at the same time. Their average test interval should be I. This is a challenge to arrange in most computerized maintenance management systems (CMMS).

Three Parallel Redundant Devices Tested Independently of Each Other

$$Pm = Pd \times Uv^3$$

$Uv = (Pm/Pd)^{1/3}$; substituting Md for Pd and Mt (tolerable) for Mm in place of Pm,

$$Uv = (Md/Mt)^{1/3}$$

$$I = 2 \times (Md/Mt)^{1/3} \times Mv$$

Again, the devices should be tested independently of each other and not at the same time, and again, this would be challenging to arrange in maintenance management systems.

Because of CMMS limitations and the likely tendency of planners and schedulers to assign all testing tasks at the same time, we need to determine I for the situation when this occurs. In this case, I is no longer an "average" of two more or less random intervals.

Now, we consider testing the devices at the same time.

n Redundant Protective Devices Tested at the Same Time: Risk-Based Criteria

Since the protective devices will each "age" at the same rate following testing, their cumulative probability of failure will climb together, and with it, unavailability.

The probability of each device failing is

$$u = 1 - e^{-\lambda t}$$

Expanding the exponential term into a power series, we get

$$u = 1 - \left(1 + (-\lambda t) + \frac{(-\lambda t)^2}{2!} + \frac{(-\lambda t)^3}{2!} + \dots \right)$$

Assuming that λ is a small number (failure rates should be very small), we can ignore the higher-order terms to arrive at the approximation

$$u = \lambda t$$

The probability of multiple devices n failing together U is

$$U = \left(1 - e^{-\lambda t} \right)^n$$

Average unavailability over the failure finding interval I will be

$$U(I) = \int_0^I \left(\left(1 - e^{-\lambda t}\right)^n \right) dt / I$$

Substituting the above approximation for u,

$$U(I) = \int_0^I \left((\lambda t)^n \right) dt / I$$

$$U(I) = (\lambda I)^n / (n+1)$$

The probability of failure of the combined system (protected plus protective) becomes

$$Pm = Pd \times (\lambda I)^n / (n+1)$$

Rearranging to solve for I,

$$I = 1/\lambda \times \left(Pm \times (n+1)/Pd \right)^{1/n}$$

Substituting $1/Mm$, $1/Md$, and $1/Mv$ for Pm, Pd, and λ, respectively, we get the general formula for testing interval for n multiple redundant devices that are tested at the same time:

$$I = Mv \left(\frac{(n+1)Md}{Mm} \right)^{1/n} \tag{A.7}$$

Note that if $n = 1$, we get the formula for a single protective device (Equation A.2):

$$I = 2 \times \frac{Md \times Mv}{Mm}$$

n Redundant Protective Devices Tested at the Same Time: Economic-Based Criteria

As before in the single protective device case, we consider the various costs, add them together for a total, and take a derivative over the variable I, then set it to zero to determine I at which total costs are minimized.

The cost of a multiple failure event is Cm (repair and downtime costs), so the cost per unit of time is Cm/Mm

Rearranging $I = Mv\,((n+1)Md/Mm)^{1/n}$ for Mm and dividing into Cm, we get

$$\frac{Cm}{Mm} = \frac{Cm\,I^{n}}{(n+1)Mv^{n}Md}$$

As before, the cost of failure finding per unit time will be

$$\frac{Cff}{I}$$

And the cost of repair of the protective system once you find a problem will be

$$\frac{Cv}{Mm}$$

Total cost becomes

$$C = \frac{Cm\,I^{n}}{(n+1)Mv^{n}Md} + \frac{Cff}{I} + \frac{Cv}{Mm}$$

Costs will rise if I is too short, because we will have more multiple failures. They will also rise if I is too frequent, because we'll be doing too many tests. Between the two extremes is a minimum point found when the slope of the cost curve is zero.

The slope of the cost curve is its derivative, $\dfrac{dC}{dt}$:

$$\frac{dC}{dt} = \frac{nCm\,I^{n-1}}{(n+1)Mv^{n}Md} - \frac{Cff}{I^{2}}$$

Setting $\dfrac{dC}{dt}$ (the slope) to zero enables us to solve for I:

$$0 = \frac{nCm\,I^{n-1}}{(n+1)Mv^{n}Md} - \frac{Cff}{I^{2}}$$

Rearranging and solving for *I*,

$$\frac{Cff}{I^2} = \frac{nCm I^{n-1}}{(n+1)Mv^n Md}$$

$$I^{(n+1)} = \frac{Cff(n+1)Mv^n Md}{nCm}$$

$$I = \left(\frac{Cff(n+1)Mv^n Md}{nCm} \right)^{\frac{1}{n+1}} \tag{A.8}$$

Complex Protective Systems

In facilities, perhaps more so than for plants and mobile equipment, we have entire protective systems for fire, detection of hazardous gases, and security. These systems can be designed for entire buildings or even entire campuses comprised of many buildings. These entire systems are protective devices that must be tested. Depending on what they are intended to protect against, the demand for them to operate is always driven by factors external to the systems themselves—for example, release of toxic gases, breakout of fire, weapons or explosives getting through checkpoints, break-in by thieves or terrorists, or breakout by inmates in prisons.

These systems will be complex combinations of series and parallel devices. They may involve multiple sensors that trigger a signal to a local panel. There may be many local panels that send another signal to a centralized panel. This may even send a signal to a remote monitor.

Consider a simple home security system. It could comprise open sensors on doors, lock closure sensors on external doors, windows, motion detectors in hallways and rooms, smoke sensors, and CO_2 sensors. They might all feed a home security panel, which in turn connects, perhaps through an internet modem or phone line, to a central monitoring station at a security company.

The testing of these systems will usually need to be done in parts, considering the functions of each part and their interconnections. For example, you can prove the functioning of the communications to the remote monitoring station, the ability of the central panel to pick up signals and transmit them, by simply triggering any of the sensing devices. Triggering the device sends a signal through an entire series of connected devices. The whole series either works or fails.

However, each sensing device and its connections must be tested individually. Each sensing "leg" is independent of the others. They operate in parallel with each other but not necessarily logically in parallel. Given that they are installed in separate areas, they don't truly back each other up as redundant devices if the detection of a problem and isolating it to a specific location is important. Each will need to be tested independently of the others. However, if all you want to know is that there is an intruder, or that there is a fire, then location in the structure is less important, because a fire in one area may not be detected by a failed smoke detector in that area, but it would likely be detected by an adjacent detector. Testing could be less frequent, and test frequencies could consider the parallel redundant nature of the arrangement.

Each design will be unique, so it is useful to draw line diagrams showing the various parallel and series combinations, determining logical testing strategies for these combinations, and calculating the failure finding test intervals for each.

Voting Systems

Another complication that can arise is when voting systems are used—for example, you need to have two of three detectors go off before you consider the situation to be in an alarm state. A simple parallel redundancy is effectively a one out of two voting logic system. These situations often arise in safety instrumented systems, as covered by IEC 61508. The standard includes methods to calculate the reliability of these systems for purposes of design. There is also a series of technical reports by the The Instrumentations Systems and Automation Society (ISA) covering safety instrumented systems, which even include tables of MTBFs for commonly used devices (ISA-TR84.00.02-2002).

If we determine the probability of failure of the voting system, Pvs, then we can use it in the general equations A.1, $I = 2Uv\,Mv$, substituting Pvs for $1/Mv$.

Mirek Generowicz presents a derivation of the equations used in the IEC standards, which has been adapted for use here. For a voting logic system,

$$Pvs = \left(\frac{N!}{(N-M+1)!(M-1)!} \right) \frac{\lambda I^{(N-M+1)}}{N-M+2} + \frac{\beta \lambda I}{2}$$

where:

    M out of N devices must work to trigger an alarm or shutdown

    λ is the failure rate of the individual devices

    β is the proportion of failures having a common cause

    I is the testing interval

    Pvs for our purposes is a probability that we'd consider tolerable (i.e., I/Mm)

The failures that have a common cause (represented by β) behave as single-channel failures in the logic. If we ignore these and deal strictly with the voting part of the system, we are left with

$$Pvs = \left(\frac{N!}{(N-M+1)!(M-1)!} \right) \frac{\lambda I^{(N-M+1)}}{N-M+2}$$

Rearranging to solve for I, substituting $1/Mv$ for λ, and substituting $1/Mm$ for Pvs,

$$I = Mv \left(\frac{(N-M+1)!(M-1)!(N-M+2)}{MmN!} \right)^{\frac{1}{N-M+1}} \tag{A.9}$$

SUMMARY OF FAILURE FINDING INTERVAL (FFI) FORMULAE

Terms Used

Cff	Cost of failure finding task
Cm	Cost of multiple failure (repair and operational costs all included)
I	Failure finding test interval
M and N	The counts of M out of N devices in a voting logic safety system
Md	MTBF of the protected function
Mi	MTBF of each component in a series of n components making up a single protective system
Mm	MTBF of the multiple failure (determined from tolerable probability of multiple failure)
Mo	MTBF of the protective device that is not caused by testing
Mv	MTBF of the protective device
n	Number of components each having Mi or number of redundant (parallel) protective devices
p	Probability of failing the protective device on test
Uv	Unavailability of the protective device

Formulae

Basic case:

$$I = 2Uv\,Mv \tag{1}$$

Single protective device (risk criteria):

$$I = 2\frac{Md \times Mv}{Mm} \tag{2}$$

If there is a probability of failing the single protective device on test,

$$I = 2\frac{Mo}{1-p}\left(\frac{Md}{Mm} - p\right) \tag{3}$$

Single protective device (cost criteria):

$$I = \left(\frac{2\,Mv\,Md\,Cff}{Cm}\right)^{1/2} \tag{4}$$

Single protective system with multiple components (risk criteria):

$$I = 2\frac{Md}{Mm}\frac{1}{\sum_{i=1}^{n}\frac{1}{Mi}} \tag{5}$$

Single protective system with multiple components (cost criteria):

$$I = \left(\frac{2 \times Md \times Cff}{\sum_{i=1}^{n}\frac{1}{Mi} \times Cm}\right)^{1/2} \tag{6}$$

Multiple redundant (parallel) protective devices (risk criteria):

$$I = Mv\left(\frac{(n+1)Md}{Mm}\right)^{1/n} \tag{7}$$

Multiple redundant (parallel) protective devices (cost criteria):

$$I = \left(\frac{Cff\,(n+1)\,Mv^n\,Md}{nCm} \right)^{\frac{1}{n+1}} \tag{8}$$

M out of N voting logic (risk criteria):

$$I = Mv \left(\frac{(N-M+1)!(M-1)!(N-M+2)}{MmN!} \right)^{\frac{1}{N-M+1}} \tag{9}$$

Appendix B: Glossary of Maintenance Terminology

TERM OR ACRONYM: DEFINITION

ACA: asset criticality analysis

Acceptable condition: that condition agreed for a particular use, not less than that demanded by statutory requirements; meeting a functional standard for equipment operation

Activity board: an information-sharing display prepared by a team or group to facilitate communication between operators and maintainers in a **total productive maintenance (TPM)** environment

Acute loss: infrequent or one-time performance shortfall, the gap between actual and optimal performance; usually associated with a major defect

Adjustments: minor tune-up actions requiring only hand tools, no parts, and usually lasting less than a half hour

AM: asset management

Apprentice: a tradesperson in training

ARB: British Air Registration Board

Area maintenance: a type of maintenance organization in which the first-line maintenance foreperson is responsible for all maintenance trades within a certain area

Asset(s): the physical resources of a business, such as plant, facilities, fleets, or their parts and components

Asset information management (AIM): the governance and management of all technical and other information related to physical assets

Asset list: a register of items usually with information on manufacturer, vendor, specifications, classification, costs, warranty, and tax status

Asset management: the systematic planning and control of a physical resource throughout its economic life

Asset number: a unique alphanumerical identification of an asset on a list, often in a database, which is used in its management

ATA: US Air Transportation Association

Autonomous maintenance: routine maintenance, PM, and PdM carried out by operators, with or without help from maintenance tradespersons, who are often part of the same team as the operators

Availability: the period of scheduled time for which an asset is capable of performing its specified function, expressed as a percentage

Available: the state of being ready for use, includes operating time and downtime for reasons other than maintenance

Backlog: work that is waiting to be done; it is estimated and awaiting planning, prioritization, scheduling, and execution

Bar code: symbols for encoding data using lines of varying thickness, designating alphanumeric characters

Benchmark: a measurable standard for high performance based on a survey or study of comparable businesses or business processes having similar key performance drivers

Benchmarking study: a formal study aimed at determining benchmarks and practices used to attain these high levels of performance

Best practice: see **successful practice**

Bill of materials (BOM): list of components, from complete assemblies to individual components and parts for an asset, usually structured in hierarchical layers from gross assemblies to minor items

Breakdown: failure of an asset to perform to a functional standard

Breakdown maintenance: a policy whereby no maintenance is done unless and until an item no longer meets its functional standard, often when the asset is no longer able to operate at all

Callback: a job that is redone because the original repair did not correct the failure

Callout: the practice of calling maintenance workers in to work at times outside of their normal workday

Capital spares: spares, usually large, expensive, difficult to obtain, or having long lead times, that are acquired as part of the capital purchase of the asset for which they are intended to be used or later, after the risk of not having them is realized; accounting often treats these spares as capital items with their value depreciated over time

Catalogue: description of a part or other stock or nonstock item that is used in the maintenance of equipment

CBM: see **condition-based maintenance** and **condition-based monitoring**

Change-out: remove a component or part and replace it with a new or rebuilt one

Charge-back: maintenance costs charged to the user department that requested the work

Chronic loss: frequently occurring performance shortfalls, the gap between actual and optimal performance

Cleaning: removing all sources of dirt, debris, and contamination for the purpose of inspection and to avoid chronic losses

CMMS: computerized maintenance management system

Code: symbolic designation, used for identification

Component: a constituent part of an asset, usually modular and replaceable, that is sometimes serialized depending on the criticality of its application and interchangeability.

Component number: designation, usually structured by system, group, or serial number

Computer, mainframe: a digital processor with the highest capacity, speed, and capability, normally used at the corporate level of a company

Computer, micro: a digital processor having moderate capability relative to a mini- or mainframe computer, usually desktop, operated by individual user

Computer, mini-: a digital processor having significant capacity but less than a mainframe, often used at the corporate or site level

Computer, workstation: equipment, usually a keyboard and display, used to access a mainframe or mini-computer; sometimes used to describe an office work area for one person; sometimes used to describe a desktop microcomputer for individual use

Conditional probability of failure: the probability that a failure will occur in a specified period, given the condition that the item has survived to the beginning of that period

Condition-based maintenance: repair or restoration of an asset based on its condition at the time; also known as on-condition maintenance

Condition-based monitoring: the monitoring of equipment performance or other condition parameters to determine the condition or "health" of the equipment or system. Condition-based monitoring is used as part of a predictive maintenance program to determine the need for condition-based maintenance

Contract maintenance: maintenance work performed by contractors

Contractor: an individual or company providing specific services to another under contract for those services, tasks, or specific results

Coordination: daily adjustment of maintenance activities to achieve the best short-term use of resources or to accommodate changes in needs for service

Corrective maintenance: maintenance done to bring an asset back to its standard functional performance

Costs, life cycle: the total cost of an item throughout its life, including design, manufacture, operation, maintenance, and disposal

Critical spares: spare parts that have high value or long lead times or are of particularly high value to the important and unspared equipment on which they are used. Unspared equipment is that which has no installed back-up or spare to take over service in event of its failure. They are carried to avoid excessive downtime in the event of a breakdown

Criticality: a measure of the importance of an asset relative to other assets

Defect: a condition that causes deviation from design or expected performance, leading to failure; a fault

Deferred maintenance: maintenance that can be or has been postponed from a schedule

Detective maintenance: testing an asset to make sure it is still functional; used primarily to check dormant devices that can fail in such a way as to be undetectable until they are needed to function

Deterioration rate: the rate at which an item approaches a departure from its functional standard

DM: see **detective maintenance**

DoD: US Department of Defense

Down: out of service, usually due to breakdown, unsatisfactory condition, or production scheduling

Downtime: the period of time during which an item is not in a condition to perform its intended function, whether scheduled or not. The distinction between "scheduled" and "unscheduled" downtime is stated where it is relevant to the discussion

EAM: enterprise asset management system

EBAM: evidence-based asset management

Emergency: a condition requiring immediate corrective action for safety, environmental, or economic risk caused by equipment breakdown

Engineering work order (EWO): a control document from engineering authorizing changes or modifications to a previous design or configuration

Equipment configuration: list of assets usually arranged to simulate process, functional, or sequential flow

Equipment repair history: a chronological list of defaults, repairs, and costs on key assets so that chronic problems can be identified and corrected, and economic decisions made

Equipment use: a measure of the accumulated hours, cycles, distance, throughput, and so on for which an asset has performed its function

ERP: enterprise resource management system

Evident failure: a failure mode that, on its own, becomes apparent to the users of the asset under normal operating circumstances

Examination: a comprehensive inspection with measurement and physical testing to determine the condition of an item

Expert system: decision support software with some ability to make or evaluate decisions based on rules or experience parameters incorporated in the database

FAA: US Federal Aviation Agency

Failure: termination of the ability of an item to perform its required function to a desired standard

Failure analysis: a study of failures to analyze the root causes, develop improvements, or eliminate or reduce the occurrence of failures

Failure coding: indexing the causes of equipment failure on which corrective action can be based, for example, lack of lubrication, operator abuse, material fatigue, and so on

Failure effect: a statement of what chain of events follows the occurrence of a failure

Failure finding task: a scheduled task used to detect whether or not an asset is in a failed state, generally used on assets that are normally dormant (e.g., safety devices, backups)

Failure mode: the event that leads to failure

Fault: see **defect**

Fault tree analysis (FTA): a top-down deductive failure analysis in which an undesirable state of a system is analyzed using Boolean logic to determine combinations of lower-level events and conditions that can lead to the undesirable state.

Five S: derived from the Japanese words *seiri* (organization), *seiton* (tidiness), *seiso* (purity), *seiketsu* (cleanliness), and *shitsuke* (discipline); focused on the workplace and successful habits that contribute to equipment condition

FMECA: failure mode, effect, and criticality analysis; a logical, progressive method used to understand the root causes of failures and their subsequent effect on production, safety, cost, quality, and so on. Used as part of RCM

Forced outage: downtime caused by a failure

Forecasting: the projection of the most probable: as in forecasting failures and maintenance activities

Functional failure: the condition of an asset not being able to fulfill a particular function at all or at the desired performance level

Functional maintenance structure: a type of maintenance organization in which the first-line maintenance foreperson is responsible for conducting a specific kind of maintenance, for example, pump maintenance, HVAC maintenance, and so on

Hard time maintenance: periodic preventive maintenance based rigidly on calendar time

Hazard and Operability Analysis (HAZOP): a structured and systematic examination of a process or operation to determine potential problems that may present risks to personnel and equipment. It is intended to identify risks design and engineering issues that may have been overlooked.

Infant mortality: failures that occur prematurely; often, these occur because of design, material, workmanship, installation, or quality problems in any work that was done prior to starting the asset up for service

Inherent capability: what an asset can initially do

Inspection: a review to determine maintenance needs, condition, and priority on equipment

Inventory: stock items that are actually on hand in a storeroom or elsewhere ready for use

Inventory control: managing the acquisition, receipt, storing, and issuance of materials and spare parts; managing the investment efficiency of the store's inventory

Inventory turnover: ratio of the value of materials and parts issues annually to the value of materials and parts on hand, expressed as a percentage

ISO: International Organization for Standardization

Issues: stock consumed through stores

Labor availability: percentage of time that the maintenance crew is free to perform productive work during a scheduled working period

Labor utilization: percentage of time that the maintenance crew is engaged in productive work during a scheduled working period

Lean manufacturing: a manufacturing system that focuses on minimizing the resources required to produce the product or service

Level of service (stores): usually measured as the ratio of stock-outs to total stores issues

Life cycle: The sequence of stages in the existence of an asset: conceptualize, plan, evaluate, design, build/procure, operate and maintain, modify, dispose

Life cycle cost (LCC): the total of all costs of the asset throughout its entire life cycle, including all work done on or to the asset, depreciation, and other costs of ownership; normally, LCC takes account of the time value of money

Logistics engineering: a systems engineering concept developed for military weapons systems; it advocates maintenance considerations in all phases of an equipment program to achieve specified reliability, maintainability, and availability requirements

Maintainability: the rapidity and ease with which maintenance operations can be performed to help prevent malfunctions or correct them if they occur, usually measured as mean time to repair (MTTR)

Maintenance: any activity carried out to retain an item in, or restore it to, an acceptable condition for use or to meet its functional standard

Maintenance audit/review: a formal review of maintenance management practices and results carried out by an independent third party for the purposes of evaluating performance and identifying areas of strength, weaknesses, and opportunities for improvement

Maintenance engineering: a staff function intended to ensure that maintenance techniques are effective, equipment is designed for optimum maintainability, persistent and chronic problems are analyzed, and corrective actions or modifications are made

Maintenance history: a record of maintenance activities and results

Maintenance policy: a principle guiding decisions for the maintenance of an asset (e.g., this asset will be run to failure and then repaired vs.

this asset will be monitored for vibrations to avoid having it fail unexpectedly)

Maintenance prevention: design of assets to avoid the need for maintenance

Maintenance route: an established route through a facility along which a maintainer carries out proactive maintenance, detective maintenance, and minor repairs on a routine basis

Maintenance schedule: a comprehensive list of planned maintenance and its sequence of occurrence based on priority in a designated period of time

Maintenance shutdown: a period of time during which a plant, department, process, or asset is removed from service specifically for maintenance

Maintenance strategy: a high-level statement of vision, mission, and objectives with a description of a general plan for achieving them; also used to describe the specific approach to be used for maintaining a specific asset

Maintenance window: the timeframe in which maintenance work can be performed without incurring any unplanned production losses

Major defect: a single defect that can cause equipment breakdown and operational losses

Margin of deterioration: the gap between the desired and the asset's inherent capability

Material safety data sheet (MSDS): an information sheet that comes with a chemical product giving the formal name of the chemical/compound, a description of its toxicity, handling instructions, warnings about its use, and first aid treatment for exposure

Menu: a selection of functional options in a software display

Meter reading: a numerical reading of the accumulated usage of an asset using an hour meter, odometer, or another device

MIL-STD: US Military Standards

Minor defect: a single defect that cannot cause losses on its own but may contribute to losses in combination with other minor defects

MRO: maintenance, repair, and overhaul; used in describing the material resource requirements to support maintenance activities

MSG: maintenance steering group

MTBF (mean time between failures): see **reliability**

MTTR (mean time to repair): see **maintainability**

Natural deterioration: the inherent deterioration that occurs in an asset as a natural result of its usage or age

NDT: nondestructive testing of equipment to detect abnormalities in physical, chemical, or electrical characteristics, using such technologies as ultrasonics (thickness), liquid dye penetrants (cracks), X-ray (weld discontinuities), and meggers (voltage generators to measure resistance). Some forms of NDT carry a risk of damaging an item or increasing the probability of failure for the item being tested (e.g., meggers and dye penetrants that require equipment disassembly), and some are completely nonintrusive (e.g., X-rays)

NES: UK Naval Engineering Standards

Nonroutine maintenance: maintenance (usually repairs) performed at irregular intervals, with each job unique, and based on inspection, failure, or condition

OEE: see **overall equipment effectiveness**

Online: the state of being available and accessible while the CMMS is operating

Opportunity maintenance: maintenance work that is performed in an unanticipated maintenance window or to take advantage of a planned maintenance window to get more work accomplished than scheduled

Outage: a term used in some industries, for example, electrical power distribution, to denote when an item or system is not in use

Outsourcing: contracting of all or a major part of the maintenance work required by an organization

Overall equipment effectiveness: OEE is a measure combining the availability, production rate (i.e., utilization of the available time), and quality rate (proportion of produced units in compliance with quality specs) of an asset

Overhaul: a comprehensive examination and restoration of an asset to an acceptable condition

Pareto: analysis to determine the minority of equipment that is causing the majority of the problems

PdM: see **predictive maintenance**

Pending work: work that has been issued for execution but is not yet completed; maintenance work in process

Performance indicators: measures that indicate the degree to which a specific function is being performed

Performance management: the act of using performance measurement as a means of identifying shortfalls and correcting them with the aim of improving overall performance results

Performance measurement: the act of measuring performance using performance indicators

Performance standard: a definition of what level of performance the user wants or needs the asset to achieve

Periodic maintenance: cyclic maintenance actions carried out at regular intervals, based on repair history data, use, or elapsed time

P-F interval: the time elapsed between potential and functional failure

Pick list: a selection of required stores items for a work order or task, normally used by stores to prepackage the needed materials for use

Plan: the comprehensive description of maintenance work to be done, including task list, parts and materials required, tools required, safety precautions to be observed, permits and other documentation requirements, an estimate of the duration of the work, effort, and costs

Planned component replacement (PCR): see **scheduled discard**

Planned maintenance: maintenance carried out according to a documented plan of tasks, skills, and resources

Planner: an individual who plans work (see **plan**); often, planners also schedule work

PM: see **preventive maintenance**

PM frequency: the frequency for performing **PM** work, also used for inspections and **PdM** and **DM** frequencies

PMO: see **preventive maintenance optimization**

Potential failure: a detectable operating or equipment condition that can be used to indicate that a failure is about to occur or in the process of occurring

Predictive maintenance (PdM): the use of measured physical parameters against known acceptable limits for detecting, analyzing, and identifying equipment problems before a failure occurs; examples include vibration analysis, sonic testing, dye testing, infrared testing, thermal testing, coolant analysis, tribology, and equipment history analysis; used to identify the need for **CBM**

Preventive maintenance (PM): maintenance carried out at predetermined intervals, or to other prescribed criteria, and intended to reduce the likelihood of a functional failure

Preventive maintenance optimization (PMO): a process of analyzing an existing PM program with the intent of optimizing its performance, sometimes used as an alternative or a complement to RCM

Priority: the relative importance of a single job in relationship to other jobs, operational needs, safety, and so on, and the time within which the job should be done; used for scheduling work orders

Proactive: a style of initiative that is anticipatory and planned for; includes PM and PdM

Process safety management (PSM): regulatory requirements designed to increase safety and environmental performance in manufacturing processes

RAM: reliability, availability and maintainability analysis

RBI: risk-based inspection,

RCFA: see **root cause failure analysis**

RCM: see **reliability centered maintenance**

RCM-R®: an optimized process for formulating failure consequence management policies for assets and processes, consisting of five pillars: data integrity, **RCM** per SAE JA1011/1012, **RAM** analysis, Weibull analysis, and continuous improvement

Reactive maintenance: maintenance repair work done as an immediate response to failure events, normally without planning, always unscheduled

Rebuild: restore an item to an acceptable condition in accordance with the original design specifications

Refurbishment: extensive work intended to restore a plant or facility to acceptable operating condition

Reliability: the ability of an item to perform a required function under stated conditions for a stated period of time; usually expressed as the mean time between failures

Reliability analysis: the process of identifying maintenance of significant items and classifying them with respect to malfunction in terms of safety, environmental, operational, and economic consequences. A possible failure mode of an item is identified, and an appropriate maintenance policy is assigned to counter it. Subsets are failure mode, effect, and criticality analysis (**FMECA**), fault tree analysis (**FTA**), risk analysis, and hazard and operability (**HAZOP**) analysis

Reliability centered maintenance (RCM): a method used to determine the appropriate failure management policies for any asset in its present operating context

Repair: to restore an item to an acceptable condition by the renewal, replacement, or mending of worn or damaged parts

Restoration: actions taken to restore an asset to its desired functional state

Return on investment (ROI): financial performance of an investment

Return on net assets (RONA): profits generated expressed as a percentage of the net value of physical assets that produced that profit

Rework: work that has to be done over

Risk assessment: the determination of a quantitative or qualitative estimate of risk related to a well-defined situation and a recognized threat

Root cause: a reason for a failure to occur. It may be related to design, installation, operation/maintenance, management, or miscellaneous matters.

Root cause failure analysis (RCFA): analysis used to determine the underlying cause or causes of a failure so that steps can be taken to manage those causes and avoid future occurrences of the failure; sometimes called root cause analysis (RCA)

Rotable: a component that is rebuilt after its useful life and rotated through maintenance stores back to use; a repairable item

Routine maintenance: see **scheduled maintenance**

Run to failure: a failure management policy that allows the asset to be run to the failed state without any effort to predict or prevent this before it occurs

Running maintenance: maintenance that can be done while the asset is in service

SAE: Society of Automotive Engineers

Schedule: a time-phased list of work to be done

Schedule compliance: the number of scheduled jobs actually accomplished during the period covered by an approved schedule; also, the number of scheduled labor hours actually worked against a planned number of scheduled labor hours, expressed as a percentage

Scheduled discard: replacement of an item at a fixed, predetermined interval, regardless of its current condition; a type of PM, a planned component replacement (**PCR**)

Scheduled maintenance: any maintenance that is prioritized to be done at a predetermined time; scheduled work may be planned or unplanned

Scheduled outage: downtime that was intended for maintenance, servicing, operational, or other purposes

Scheduled restoration: repair or restoration of an asset at a predetermined interval, regardless of its current condition; a type of **PM**

Scheduler: an individual who schedules work; see **planner**

Scheduling cycle: the length of time for which scheduling is normally done for work backlog, often weekly or biweekly

Scoping: outlining the extent and detail of work to be done and the resources needed

Seasonal maintenance: maintenance work carried out at a specific time of year; for example, repair of potholed roads in northern climates, repairs to school buildings during vacation periods

Service level: an expression of the percentage of spares that are issued on demand; also, a specification of the desired service standards to be met by a contractor

Servicing: the replenishment of consumables needed to keep an item in operating condition (e.g., lube oil, ink, wearing surfaces, cleaning of working surfaces)

Setup and adjustment: a process of changing from one manufacturing configuration to another to accommodate a change in product being produced on the same asset

Shelf life: that period of time during which materials in storage remain in an acceptable condition

Shutdown: that period of time when equipment is out of service; also refers to major maintenance work, in which primary producing assets are down while the maintenance is being performed

Shutdown maintenance: maintenance done while the asset is out of service, as in the annual plant shutdown

Six losses: in TPM, these are the major losses that occur due to inadequate equipment operation or condition: breakdown; setup and adjustment; minor stoppages; speed reductions; quality defects and rework; and yield reductions

Specifications: physical, chemical, or performance characteristics of equipment, parts, or work required to meet minimum acceptable standards

Sporadic loss: see **acute loss**

Standard job: a preplanned maintenance job with all details required for work execution delineated and stored (usually in the **CMMS**, **EAM**, or **ERP**) for repeated use

Standby: assets that are used as backups to others, that are installed or available but not in use

Standing work order (SWO): a work order that remains open, usually for the annual budget cycle, to accommodate information on small jobs or for specific tasks

Stock: a term used to describe parts that are normally kept on hand in the storeroom

Strategy: 1. the overall approach for managing the life cycle of a specific physical asset (e.g., its maintenance strategy); 2. an overall direction and flexible high-level plan for business

Successful practice: a practice that leads to superior performance or results in a specific process; sometimes called "best practice," but this implies that it is the only way to execute the practice

Superintendent: a second-line manager who is responsible for a maintenance group or department

Supervisor: a first-line manager who is responsible for a group of tradespersons

Survey: a formal inspection of a plant, facility, civil infrastructure, or vehicle to look for condition and defects

Tactics: the choices made to implement a strategy and manage the people, processes, and physical asset infrastructure that make up your business

Task: a single item on a task list that informs an inspector or maintainer what to do; an instruction

Task list: directions to an inspector or maintainer telling him or her what to do and in what sequence; for example, check oil level, clean, adjust, lubricate, replace, and so on

Terotechnology: an integration of management, financial, engineering, operating maintenance, and other practices applied to physical assets in pursuit of an economical life cycle

Total productive maintenance (TPM): companywide equipment management program emphasizing operator involvement in equipment maintenance and continuous improvement in equipment effectiveness

Trade: a specific skill or set of related skills in a particular area (e.g., millwright, electrician, machinist, boilermaker, carpenter, rigger, etc.)

Tradesperson: skilled worker who normally has completed an apprenticeship program; in some jurisdictions, certain tradespersons must be tested and licensed in their respective trades

Unplanned maintenance: maintenance done without planning; could be related to a breakdown, running repair, or corrective work; unplanned maintenance may be scheduled during the normal schedule cycle

Up: used in reference to an asset that is available and being used

Uptime: the period of time during which an item is in a condition to perform its intended function, whether it is in use or not

Utilization factor: usage of an asset expressed as a percentage of schedule time

Variance analysis: interpretation of the causes of a difference between actual and some norm, budget, or estimate

Visual control: the use of easy-to-read indicators to show equipment status and performance (e.g., red, yellow, or green gauge markings; normal reading zone indicators; color-coded oil cans and filler caps)

Warranty: coverage for repair costs incurred in the event of a defect caused by a supplier of equipment, materials, or services

Weibull analysis: a statistical analysis used for life data analysis, among other applications, consisting of plotting operating time at failure versus percentage of accumulated failures on log-scale paper

Work in process (WIP): partially completed production "product" at some interim stage in the production process; product that is still being worked on prior to being considered ready to deliver

Work order (WO): a unique control document that comprehensively describes the job to be done; may include formal requisition for maintenance, authorization, and charge codes, as well as a record of what work was actually done, time, and materials used

Work request (WR): a simple request for maintenance service or work requiring no planning or scheduling but usually a statement of the problem; usually precedes the issuance of a work order

Workload: the number of labor hours needed to carry out a maintenance program, including all scheduled and unscheduled work and maintenance support of project work

Epilogue

We've seen how reliability centered maintenance originated in the aircraft industry, where high reliability is a must-have to achieve flight safety at a reasonable cost. In the early days of commercial aviation, mankind, just getting used to this new and fast mode of transportation, paid a high price in terms of maintenance man-hours and lives. If the industry was to grow profitably, it had to match its maintenance efforts to the nature of the failures that were occurring. Early attempts at decision logic helped to a limited degree, but resulted in programs that would render the fledgling industry uneconomic and unprofitable. Worst still, there seemed to be a point of diminishing returns: the more maintenance they performed, the less reliable the aircraft systems became. Something was wrong. At United Aircraft in the 1970s, Stan Nowlan and Howard Heap carried out their landmark study, "Reliability-Centered Maintenance,"* that would turn that situation around. Matching maintenance activities to failure causes worked well, changing the type of work being done, reducing its costs, and increasing flight safety. Their work is the foundation on which all subsequent RCM methods are based, including RCM-R®.

Military organizations and the nuclear power industry were quick to adapt the original work to their purposes. Commercial variants began to appear, and in the early 1990s two dominant commercial variations appeared. By the end of the decade, the US military wanted a commercial standard that could replace the need for its complex and cumbersome military RCM standards and handbooks. At the request of the US Department of Defense, the Society of Automotive Engineers (SAE) was asked to generate such a standard. Various industry experts contributed to the standard writing effort, and the results was SAE JA1011†. The military had what it needed to reduce its future procurement costs, and industry had a standard to help define and clear up confusion among what was becoming a crowded competitive marketplace for RCM methods and services. A few years later, SAE also produced a set of guidelines for application of

---

* Nowlan, F. Stanley, and Howard F. Heap, "Reliability-Centered Maintenance," Department of Defense, Washington, DC, 1978. Report number AD-A066579
† SAE JA1011, "Evaluation Criteria for Reliability-Centered Maintenance (RCM) Processes", Aug 1999

JA1011, known as SAE JA1012. Since then, little has changed in the RCM domain, and industry applied the method with mixed results.

The authors both worked with RCM during this period and, in addition to gaining expertise with RCM, learned a number of useful things about how it should and should not be implemented. Independent of each other, one adapted technical analysis to enhance the method and the other adapted training and delivery methods to a market that was evolving. As economies struggled with the Great Recession, companies became more and more lean (some perhaps too lean), and demographics shifted the composition of the workforce, impacting on the level of experience that can be found in companies today. When Jesus and James met in 2015 and discussed RCM, they realized that their combined insights could be of great value to this reshaped marketplace that had needs extending beyond those that existed when RCM was first created and during its early formative years. RCM-R® began with Jesus' work on the technical enhancement of RCM and its alignment to international standards. That was further enhanced with James' insights into delivery methods. They decided to pool their knowledge and experience and produce a new book on this newly re-engineered approach to be known as "Reliability Centered Maintenance-Reengineered: RCM-R®."

For the next year and a half, they collaborated on this book you have just finished reading. In it they describe the early history of RCM and its development into the successful method it can be, if applied correctly. There is a clear and substantial business case for using RCM, based on creating the most cost-effective failure management program you can while enhancing systems' performance. This results in safer and more environmentally friendly operations, more dependable production and services delivery, and excellent management of risks—not only dealing with them when they occur, but actually reducing their occurrence.

RCM-R® goes beyond what RCM alone can do. The basic successful method as defined in SAE JA1011 remains intact. RCM-R® enhances that method, linking it to international standards for risk management and adding a degree of technical rigor rarely seen outside of the military, nuclear, and aircraft industries. It adds a great deal of emphasis on what it takes to implement the method successfully—not only as a project (as has so often been done with other RCM methods), but as a sustainable program, and on leveraging the analysis results to maximize value generation and align closely with the intentions and precepts of the new international standard for asset management, ISO 55001*.

Asset management, as it is now defined, is the coordinated effort of an organization to realize value from assets. This requires the balancing of costs, risks, opportunities, and performance, and it is normally implemented through an asset management system (not to be confused with purely computer based "systems"). Achieving the requirements of this new standard entails an effort that must be based in part on risk. A method to identify risks arising from those assets is needed, and RCM-R® is just such a method. It goes beyond other RCM methods in its emphasis on technical rigor, implementation, and leveraging of its findings. It truly engages the entire organization to realize value from its assets over their entire life cycle, from conceptual design to decommissioning and disposal.

RCM, as defined in SAE JA1011, has seen little development since the standard first emerged. Commercial interests have dominated the RCM landscape since the mid-1990s. Since the mid-2000s, the original thought leaders have either passed away or faded into obscurity. The time is ripe for a new perspective and tool for reliability, more closely aligned with today's thinking on asset management, more effective in a harsh business climate, leveraging today's technological advancements and accommodating the shifting demographics in our dynamic workforce that is actually getting younger. It is time for RCM to be re-engineered. We hope that our blend of experience and experimentation, ultimately culminating in our collaboration on the writing of this book, provides that fresh approach. We hope that the work we've put into developing our methods and documenting them here contributes to safety, environmental integrity, risk reduction, and profitable industrial capability for years to come.

Thank you for reading our work. We hope you will gain sufficient faith in what we have done to employ what we have learned in helping your organization on that journey of excellence we know as asset management, enabled and enhanced by RCM-R®.

Jesús and James

---

* "Asset management—Management systems—Requirements", ISO 55001, Jan 2015, International Standards Organization.

Index

Milton Keynes UK
Ingram Content Group UK Ltd.
UKHW051535141024
449569UK00001B/41